Algae in Agrobiology

*Series Editors
Jack Legrand and Gilles Trystram*

Algae in Agrobiology

Realities and Perspectives

Joël Fleurence

   WILEY

First published 2023 in Great Britain and the United States by ISTE Ltd and John Wiley & Sons, Inc.

Apart from any fair dealing for the purposes of research or private study, or criticism or review, as permitted under the Copyright, Designs and Patents Act 1988, this publication may only be reproduced, stored or transmitted, in any form or by any means, with the prior permission in writing of the publishers, or in the case of reprographic reproduction in accordance with the terms and licenses issued by the CLA. Enquiries concerning reproduction outside these terms should be sent to the publishers at the undermentioned address:

ISTE Ltd
27-37 St George's Road
London SW19 4EU
UK

www.iste.co.uk

John Wiley & Sons, Inc.
111 River Street
Hoboken, NJ 07030
USA

www.wiley.com

© ISTE Ltd 2023

The rights of Joël Fleurence to be identified as the author of this work have been asserted by him in accordance with the Copyright, Designs and Patents Act 1988.

Any opinions, findings, and conclusions or recommendations expressed in this material are those of the author(s), contributor(s) or editor(s) and do not necessarily reflect the views of ISTE Group.

Library of Congress Control Number: 2023938461

British Library Cataloguing-in-Publication Data
A CIP record for this book is available from the British Library
ISBN 978-1-78630-919-8

Contents

Preface vii

Introduction ix

Chapter 1. History 1

    1.1. The different types of kelp 1
    1.2. Historical applications 3

Chapter 2. Traditional Applications of Algae in the Cultivation Plants ... 11

    2.1. Uses for soil amendment 11
    2.2. Soil fertilization 14
    2.3. Improvement of composts for agricultural use 16

Chapter 3. Biostimulation Activities on Plant Productions ... 23

    3.1. Stimulation of growth 25
    3.2. Tolerance to water stress 38
    3.3. Tolerance to salt stress 45
    3.4. Tolerance to thermal stress 58
    3.5. The quality of the products 60

Chapter 4. Feeding of Livestock 69

    4.1. Ruminant nutrition 69
    4.2. Pig nutrition 80
    4.3. Horse nutrition 87
    4.4. Poultry nutrition 90
    4.5. Nutrition of rabbits 100
    4.6. Nutrition of animals produced by aquaculture 103

4.6.1. Fish	104
4.6.2. Mollusks	118
4.6.3. Crustaceans	124
4.6.4. Echinoderms	128

Chapter 5. The Biological Activities of Algae in Plant or Animal Health . . 135

5.1. Antiparasitic and antimicrobial activities	135
5.1.1. Plant parasites and pathogens	135
5.1.2. Animal parasites and pathogens	146
5.2. Induction of plant defense mechanisms	151
5.2.1. The hypersensitivity reaction	151
5.2.2. Other mechanisms	154
5.3. Activation of the immune system	158
5.3.1. The case of fish raised by aquaculture	158
5.3.2. Other aquaculture animals	163
5.3.3. The case of terrestrial livestock	165

Conclusion	169
References	171
Index	183

Preface

The use of algae is an ancient practice in agriculture, carried out on the coasts of many countries. The algae resource, resulting from beaching, has sometimes been mixed with sand, or even manure, to constitute a humus suitable for crops. This is the case with the surface layer of soil found on the island of Aran located off the west coast of Ireland. Traditionally, algae known as "kelp" or "wrack" are used for soil amendment or fertilization. They are also integrated in the breeding process through their nutritional contribution in animal feed. This book reviews the traditional and current algae applications in agriculture and livestock breeding. The latter aspect is dealt with although it does not meet the strict definition of *agri cultura* (field culture). The implication of livestock in agricultural work and in the supply of food proteins has indeed contributed to the undeniable success of agriculture.

The role played by the use of algae in the development of food and vegetable crops on the European coasts (Ireland, United Kingdom, France) or in the feed of livestock (sheep, cattle, horses) is initially developed as a historical preamble to the book.

A report on the activities of agronomic or veterinary interest from macroalgae or cyanobacteria is proposed. In particular, it focuses on the description of hormones and algal oligosaccharides involved respectively in plant growth, tolerance to abiotic stresses (thermal shock, hydric stress, saline stress) and resistance to abiotic stresses (heat shock, water stress, salt stress) and biotic stresses (viral, bacterial and fungal infections).

This book also deals with the effect of algae, whether macro- or microalgae, on zootechnical performance (growth) and the health of livestock. It reports on the immunostimulant properties of polysaccharides or algal extracts in livestock or species of aquaculture interest (salmon, sea bream).

Used since time immemorial in agriculture, algae are inputs of biological origin and thus meet the expectations of a more ecological production activity. The agronomic studies associated with them fully meet the agrobiology, which is defined as "all biological research applied to agriculture".

Another definition of agrobiology is "the whole of the agricultural techniques which aim to respect nature through the return to ancestral practices".

Other definitions linking agrobiology to organic or reasoned agriculture are also frequently put forward.

But whatever the definition, the use of algae in agriculture is in line with the objectives of agrobiology. They also meet the expectations of the closely related sector of agriculture, which is the sustainable breeding of livestock, whether terrestrial or aquatic.

May 2023

Introduction

Algae are photosynthetic eukaryotic organisms that live mainly in aquatic environments. A distinction is made between unicellular algae or microalgae and multicellular algae or macroalgae. Microalgae are organisms that can be found in marine, brackish and fresh waters, as well as in humid air such as the atmosphere, in the form of aerosols, or on building facades (Fleurence 2021a). Macroalgae represent the resource traditionally used in agriculture, whether for improving crop yields or for breeding. Microalgae or sludge containing them are sometimes tested as biofertilizers for many crops. For nearly 70 years, microalgae have been used in the nutrition of animals produced by aquaculture (fish, shellfish, mollusks).

Algae are classically classified in botanical groups on the basis of their pigmentary composition (see Table I.1). With regard to phylogenetic systematics, this classification appears obsolete today and is therefore less applied by scientists. On the other hand, it is kept by naturalists, some biologists and professionals involved in the valorization of algae, because it proves to be very useful for taxonomic sorting of algae in the field.

In classical systematics, three major groups of algae or phyla distinguishing themselves by their pigment content have been established, namely Chlorophytes (green algae), Rhodophytes (red algae) and Chromophytes (brown algae) (see Figure I.1).

Phyla are organized into classes, the main ones of which are listed in Table I.2.

Branch	Main pigment	Secondary pigments
Chlorophytes or green algae	Chlorophyll a	Chlorophyll b
Rhodophytes or red algae	Chlorophyll a	– Phycoerythrin – Phycoerythrocyanin – Allophycocyanin – Chlorophyll d
Chromophytes or brown algae	Chlorophyll a	– Chlorophyll c – Excess of carotenoids (carotene, xanthophyll, fucoxanthin)

Table I.1. *Classification of algae according to their pigment composition*

Branch	Main classes
Chlorophytes (green algae)	– Chlorophyceae – Ulvophyceae
Rhodophytes (red algae)	Rhodophyceae
Chromophytes (brown algae)	Pheophyceae

Table I.2. *Major classes of macroalgae associated with the phyla Chlorophytes, Rhodophytes and Chromophytes (from Dawes 2016)*

This dichotomy based on pigment composition has been enriched by the application of the endosymbiotic theory to algae. According to this theory, it is now necessary to distinguish the phyla of Chlorobiontes (green algae), Rhodobiontes (red algae) and Chrysobiontes (brown algae) (Perez 1997, pp. 11–64).

For practical reasons related to the uses of algae, only the traditional classification previously described (see Table I.1) will be used as a basis of reference.

The branch Chlorophyta or green algae includes nearly 6,000 species occurring as micro- or macroalgae. The majority of these species live in fresh water (90%) and only 10% in marine waters (Dawes 2016). Among the latter are species belonging to the genus *Ulva* which are marine algae living on the top of the foreshore and easily accessible at low tide (see Figure I.2).

Figure I.1. *Photos of some algae on the foreshore belonging to Chlorophytes (a, b), Rhodophytes (c) and Chromophytes (d, e): a)* Ulva *sp.* b) Enteromorpha *sp. (green algae). c)* Palmaria palmata *(red algae). d)* Fucus *sp. and* Ascophyllum nodosum. *e)* Laminaria *sp. (brown algae) (photo credits © J. Fleurence, 2010, 2019). For a color version of this figure, see www.iste.co.uk/fleurence/algae.zip*

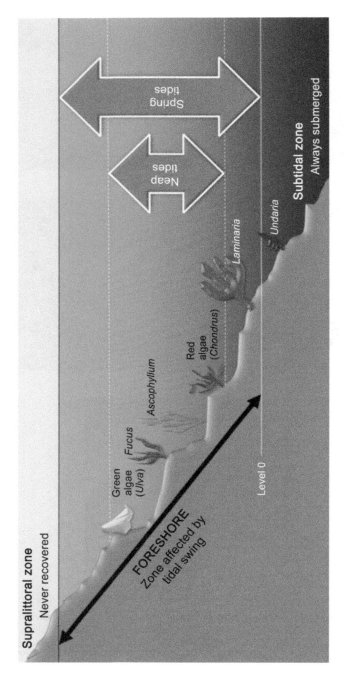

Figure I.2. *Bathymetric distribution of some algae, for example, kelp, valued in agriculture (source: Y.-F. Pouchus after Fleurence (2018)). For a color version of this figure, see www.iste.co.uk/fleurence/algae.zip*

Rhodophytes are considered one of the oldest groups of eukaryotic algae (1.7 billion years old) (Perez 1997, pp. 11–64; Baweja et al. 2016). Some species such as *Lithothamnium calcareum* (maerl) are intensively exploited for agricultural needs related to soil amendment (see section 2.1).

Chromophytes or brown algae are characterized by a very large morphological diversity. They constitute the group mainly represented in the subpolar and equatorial regions. The brown algae of the class Pheophyceae are the subject of many valuations, whether in the agricultural, food, pharmaceutical or cosmetic fields. The main species concerned are macroalgae of the order Fucales (*Ascophyllum* sp., *Fucus* sp.) or Laminariales (*Laminaria* sp., *Undaria* sp.). The Fucales are located towards the top of the foreshore and are easily accessible via on-foot fishing, which is not the case for the Laminariales, which generally grow at the limit of the subtidal zone (see Figure I.2). The species *Ascophyllum nodosum* and *Fucus vesiculosus* are the main species that make up the mixture of algae known to the public as kelp or wrack. These algae were harvested by coastal populations for many agricultural uses such as crop fertilization or livestock feed.

In addition to the ancient applications of algae in soil amendment and crop fertilization, algae is now used as a biostimulant for plant production and as an input for animal nutrition. These uses are justified by the presence in algae of original and varied molecules such as phytohormones, elicitors of defense mechanisms to biotic stresses and other substances involved in the physiology of plant or animal nutrition. All these activities of agronomic or zootechnical interest are at the origin of the current algae craze in agrobiology.

1
History

1.1. The different types of kelp

Seaweed has often been used by coastal populations to improve the physical structure of the soil or to provide nutrients for flower beds. These are often algae washed up on beaches by storms and called wrecked kelp (see Figure 1.1(a)). The latter is very distinct from shoreline kelp (see Figure 1.1(b)) which is present on the foreshore and which is manually harvested during low tides. In addition to these two categories, there is also ground kelp which is harvested by dredging the bottom with boats (see section 2.1). Kelp has often been used as a soil amendment for crops. Depending on the region and time, the algal resource exploited for agricultural purposes has been given various and very different names (see Table 1.1).

Country or region	Old name	Traditional name
Brittany	*Gouesmon*	*Goémon*
Normandy	*Vraicq* or *Vraic*	*Varech*
Island of Oleron	*Gaymon*	*Sart*
Channel Islands	Vraic	Varech
United Kingdom	Wrack	Kelp

Table 1.1. *Examples of the names of the algal resource used historically in agriculture according to regions and the time period (from Lami 1941, Blench 1966 and Desouches 1972)*

The purpose of introducing algae into the soil is to change the composition and texture of the soil. This practice of soil preparation was used by the first inhabitants of the island of Aran off the coast of Galway (Ireland) to create a layer of humus necessary for the establishment of food crops (potatoes). This practice has shaped the landscape of the island, which appears as a succession of plots protected from ocean winds and dedicated to potato cultivation (see Figure 1.2). The stranded algae is preliminarily mixed with sand and sometimes with manure before being spread in the soil.

a) b)

Figure 1.1. *Wrecked kelp (a) and shoreline kelp (b) (photo credits © J. Fleurence, 2010, 2020). For a color version of this figure, see www.iste.co.uk/fleurence/algae.zip*

This process of amendment is notably evoked in the 1934 documentary film by Robert J. Flaherty, *Man of Aran*[1].

Figure 1.2. *Former plots dedicated to food crops on the island of Aran (photo credit © J. Fleurence, 2010). For a color version of this figure, see www.iste.co.uk/fleurence/algae.zip*

---

1 See http://www.film-documentaire.fr/4DACTION/w_fiche_film/6443.

1.2. Historical applications

The direct use of algae or composted algae for crops is often related to a need for soil amendment and crop fertilization. Seaweed manure was used as early as the 1st century CE for growing cabbage (Craigie 2011). The Roman writer Columella recommended that cabbage plants in the six-leaf stage of development be contacted at the root level with seaweed manure, also used to mulch crops. Another Roman, Palladius, in the 4th century CE, also suggested the early spring application of algal manure to the roots of pomegranate and lemon trees (Arzel 1994; Craigie 2011).

On the English speaking island of Jersey, algae have been used in agriculture since the 12th century (Blench 1966). The algal resource used is referred to as "vraic" or "wrack", which appears to be a distortion of the French word *varech* or the Old English *wraec* (Blench 1966). The use of this resource by farmers was regulated by a Code of Laws, published in 1771 and supervised by two sworn officers. This code defines, among other things, the number of vraic batches that can be allocated according to the owner's cultivable land area (see Table 1.2).

Cultivable area	Number of vraic batches
More than 60 borders	6
45–60 borders	5
30–45 borders	4
18–30 borders	3
8–18 borders	2
3–8 borders	1

Table 1.2. *Allocation of batches of kelp or wrack (vraic) according to the 1711 Code of Laws of the island of Jersey (from Blench (1966))*

In the 19th century, algae were the most important fertilizing agent in the island's cultural practices. They were mainly used in the form of ash (Quayle 1815) and their use was subject to the seasonality of the crop. In the case of wheat, the vraic ash was spread on the soil before winter plowing (November–December). Summer vraic ash was particularly valuable for wheat cultivation, and many coastal residents increased their income by harvesting drift seaweed and burning it to ash. Until the middle of the 19th century, varieties of algae constituting vraic had not been identified. In 1860, Dally classified the algae used in the process into two categories: "sawn" or cut "vraic", and "growing vraic" or washed up vraic (see Table 1.3, Blench 1966). For each of these categories, he established a taxonomic identification of the species concerned. The species making up the sawn bulk were mainly pale fucales belonging to the genera *Ascophyllum* and *Fucus*. On the other hand, the algae constituting the washed up vraic were brown algae of the order Laminariales or red algae such as *Fucus palmatus* (see Tables 1.3 and 1.5).

Sawn vraic (wrack or shoreline kelp)	Washed up vraic (wrack or wrecked kelp)
F. nodosus (*Ascophyllum nodosum*)	*F. lareus* (*Himanthalia elongata*)
F. vesiculosus	*F. saccharina* (*Laminaria saccharina*)
F. canaliculatus (*Pelvetia canaliculata*)	*F. digitatus* (*Laminaria digitata*)
F. serratus	*F. palmatus* (*Rhodymenia palmata**)

* *Rhodymenia palmata*; see Table 1.5.

Table 1.3. *Algal species constituting the sawn vraic and growing vraic according to Dally (1860) (from Blench (1966))*

In France and more particularly in Brittany, kelp has also been used for centuries in agriculture. In 1681, Colbert published a royal decree to regulate the harvesting of shoreline kelp (see article I) and from wrecks (see article V) with the following articles:

> Article I "The inhabitants of the parishes situated on the coasts of the sea, will assemble on the first Sunday of January of each year, at the end of the parish mass, to regulate the days on which the cutting of the grass called kelp or vraicq, sart or gouesmon, growing in the sea in the place of their territory, must begin or finish."
>
> Article V "Nevertheless, let us allow any person to take indifferently at any time and in any place, the vraicq thrown by the waves on the shores and to transport them wherever they want" (Desouches 1972).

In the 19th century, wrecked kelp was a resource often used as fertilizer by coastal populations, whether they were islands or not. The nature of this kelp depended on the algal flora, the topography of the coasts and the strength of the waves, in other words the sheltered or beaten mode which applies to the coastal strip. On the Breton coasts, the wrecked kelp exploited by the populations was mainly composed of *Laminaria*, *Fucus* and *Ascophyllum*. On the other hand, in certain bays with a muddy inlet (e.g. in Belfast Bay), green algae of the genus *Ulva* constitute, almost by themselves, this type of kelp (Sauvageau 1920). The use of kelp on crops by coastal populations also had a phytosanitary advantage at a time when farming practices could not rely on agrochemicals. Indeed, algal manure, unlike farmyard manure, is free of weed seeds, pathogenic fungi and phytophagous insect larvae. This empirical knowledge of the benefits of using algae as a natural crop protection fertilizer has contributed to the development of many market gardening areas along the European coast. However, the preferred use of wrecked or shoreline kelp may vary by region (see Table 1.4).

Geographical sites	Wrecked kelp	Shoreline kelp	Species	Types of crops
Belfast Bay	++++	–	– *Ulva latissima* – *Monostroma* sp. – *Enteromorpha* sp.	Potatoes
Island of Ré	++++	++	*Fucus vesiculosus*	– Vegetables – Barley – Wheat – Vine
Island of Noirmoutier	–	++++	*Rytiphlaea pinastroides*	
Islands of Ouessant, Molène, Batz, Sein	++++	–	*Laminaria cloustonii*	Potatoes
Roscoff	++++	++++	– *Fucus vesiculosus* or *serratus* – *Ascophyllum nodosum* – *Laminaria* sp. – *Himanthalia elongata* (Fillit)	– Cauliflower – Onions – Potatoes – Artichokes – Parsnips
Alaskan Shores	++++	–	*Alaria fistulosa*	Potatoes
Basque Coast	++++	–	– *Saccorhiza bulbosa* – *Cystoseira* sp. – *Calliblepharis ciliata*	Not specified
Breton coasts (except Roscoff)	++++	+	– *Laminaria* sp. – *Fucus serratus*	
Danish Coasts	++++	–	Not specified	Potatoes
Irish Shores	++++	–	– *Fucus vesiculosus* – *Ascophyllum nodosum* – *Laminaria* sp.	Potatoes
Japanese coasts	++++	–	Not specified	Rice

Table 1.4. *Type of kelp and constituent species used in different regions of Europe in the 19th and 20th centuries for agricultural activities (from Sauvageau 1920) (the names of the algae are those that prevailed at the beginning of the 20th century; some species may have changed names; refer to https://www.algaebase.org/)*

On the island of Ré, ground kelp, mainly composed of *Fucus*, is widespread in layers on barley seedlings. The effect produced is spectacular, and the barley of the island is very appreciated by the brewery industry (Sauvageau 1920). Shoreline kelp or cut kelp is the object of a particular treatment. It is deposited in regular layers with fresh horse manure and left in the open air until the end of November. The manure obtained is then used for the preparation of cereal crops and vineyards.

The use of kelp on the island of Ré also concerns expensive market gardening, and the vegetables obtained are of excellent quality. No gustatory impact linked to the use of the marine fertilizer is to be deplored. On the other hand, the wine resulting from the culture of the vine with this type of fertilizer presents a pronounced marine taste, thus decreasing the market value of the product.

On the island of Noirmoutier in the 1860s, kelp was used to the exclusion of any other fertilizer for all crops. This resource known as red seaweed is considered by farmers as a very powerful fertilizer. The seaweed that composes it is *Rytiphlaea pinastroides*, a red seaweed of rather limited abundance.

Figure 1.3. *Harvesting kelp on the Breton coast at the end of the 19th century (source: illustration by F. Gillot, photo credit © J. Fleurence, 2022). For a color version of this figure, see www.iste.co.uk/fleurence/algae.zip*

The use of algae as fertilizer is an ancient practice in the Roscoff region. This use was at the origin, in the 19th and 20th centuries, of the exponential development of the market gardening activity in this basin of production. In Roscoff, the shoreline and wrecked kelp were indifferently used as fertilizer. The shoreline kelp was mainly composed of *Fucus serratus,* and the wrecked kelp was composed of Fucales and various Laminariales (see Table 1.4). In particular, wrecked kelp was very abundant after a storm and in such circumstances that shore dwellers hastily left their occupations to collect and load the valuable sea fertilizer onto carts (see Figure 1.3).

On a more ad hoc basis, the brown seaweed *Himanthalia elongata* was harvested as shoreline kelp by farmers during the fall equinox. This seaweed locally called "Fillit" was, after draining, spread in the artichoke fields.

The algal resource, when spread fresh after draining, is referred to as green kelp (Sauvageau 1920). This type of kelp is excellent for potato production, but appears to be unsuitable for onion cultivation. More generally, farmers prefer to use partially fermented kelp before it is incorporated into crops.

In Roscoff, the kelp used for many market garden crops of economic interest (see Table 1.3) was the source of the prosperity of the Léon region until the middle of the 20th century.

In the region of Morlaix or Concarneau, farmers also used a particular kelp made of a calcareous alga, *Lithothamnium calcareum* or maerl for the amendment of cultivable grounds. This alga, which lives in the subtidal zone, could be pulled up by the currents and run aground in areas that were still emerged or be recovered by dredging operations at shallow depths via boats.

On the Normandy coasts, the inhabitants have the habit of collecting the old blades of *Laminaria hyperborea* (formerly *L. cloustonii*) (see Table 1.5) which detach from the thallus in spring and wash up on the beaches. This kelp, called "mantelet" by the Normans and "April kelp" by the Bretons, is used as a fertilizer for vegetable crops, apple, pear and peach orchards (Sauvageau 1920; Perez 1997, pp. 11–64).

On the Breton islands of Batz, Molène, Sein and Ouessant, the stipes of beached *Laminaria (Laminaria cloustonii)* are burned and the ashes are used locally or on the mainland for crop fertilization and more particularly that of the potato.

Historical uses of kelp for soil amendment and fertilization show that this resource can be used directly after draining (green kelp), after drying and partial decomposition (manure), or after burning (ash). The last process is practical for transporting this marine fertilizer over longer distances than those corresponding to the coastal strip.

The use of algae in agriculture has not historically been limited to plant productions. It has also been applied to the breeding of livestock, whether they are used for field work (cattle, horses) or for the production of meat.

In the treatise *Bellum Africanum* written in 45 BCE, it is reported that the Greeks in times of famine proceeded to harvest shoreline algae and, after rinsing, used the algae to feed their cows (Evans and Critchley 2014). In the text of the Icelandic sagas, examples of algae consumption by livestock are also cited. For example, during dearth periods, algae was eaten on beaches by sheep or collected by natives to feed sheep, horses or cattle for six to eight weeks per year (Makkar et al. 2016).

In the 18th century, the consumption of algae by livestock in the northern regions of Europe is well documented. In Scandinavia, for example, sheep and goats actively sought out *Palmaria palmata* (formerly *Rhodymenia palmata*) (see Table 1.5) for consumption. This behavior led the botanist Gunner to ironically name this species "*Fucus ovinus*" (Sauvageau 1920). However, it should be noted that the species *F. ovinus* is today listed in Algaebase as *F. vesiculosus*, which distances it from the description reported by Sauvageau (1920). On the Swedish island of Gotland (formerly Gothland), pig feeding relies on the supply of a preparation based on boiled *Fucus vesiculosus*, mixed with a coarse meal. In Norway, *F. vesiculosus* and *Chorda filum* are used as fodder for cattle feed.

In the 19th and 20th centuries in northern European countries, including France, kelp was used occasionally or systematically to feed ruminants (cows, calves, pigs). This practice was the result of the observation of the animals' behavior towards the kelp washed up on the banks. Thus, in many Breton regions (Roscoff, Batz Island), cows willingly consume the species *Palmaria palmata* washed up on the shore. This species is elsewhere called "cow kelp" or "cattle kelp" by the local populations.

On the Scottish island of Lewis, cattle and sheep are allowed to roam freely and feed on the algae present on the shores. Sheep prefer *P. palmata* to the detriment of the brown alga *Alaria esculenta* although it is present in abundance. Cows, on the other hand, selectively consume brown algae of the genera *Alaria* or *Laminaria* (Sauvageau 1920).

During World War I, some countries used the algal resource to feed livestock. In Norway, seaweed was frequently fed to livestock after desalting by rinsing and boiling. These were mainly brown algae such as *Alaria escu lanta* called "Kutara" (cow alga) or *Ascophyllum nodosum* called "Grisetang" (pig alga) (Sauvageau 1920).

In Germany, algae were integrated in the food of pigs, cows, ducks and sheep. In France, the army proceeded in 1917 to experiment with feeding horses brown algae belonging to the genera *Laminaria*, *Saccharina* and *Fucus*. These experiments, which followed the shortages of oats and fodder, were carried out on cull animals. Carried out over several weeks, they showed the harmless nature of the consumption of algae by the equids and in a certain number of cases a weight gain of about 6% (Sauvageau 1920).

However, after this world conflict, a gradual consensus was established that the contribution of algae in the nutrition of livestock was less than that of other plant sources. It was not until the second half of the 20th century and the advances in zootechnical and biochemical research that the use of algae in animal feed was revived.

Former name	Current name
***Fucus** palmatus* *Rhodymenia palmata*	*Palmaria palmata*
Laminaria cloustonii	*Laminaria hyperborea*
Fucus saccharinus *Laminaria saccharina*	*Saccharina latissima*
Fucus digitatus	*Laminaria digitata*
Fucus ovinus	*Palmaria palmata* (?) *Fucus vesiculosus*
***Fucus** canaliculatus*	*Pelvetia canaliculata*

Table 1.5. *Correspondences between old species names and current names (from Sauvageau 1920, Blench 1966 and Algaebase 2021)*

2

Traditional Applications of Algae in the Cultivation Plants

2.1. Uses for soil amendment

The purpose of soil amendment is to improve the physical or chemical structure of the soil to prepare it for cultivation. Algae, rich in polysaccharides retaining water (alginates, agar, carrageenans) or minerals such as calcium carbonate (limestone), have structuring properties capable of modifying the texture of agricultural land. The addition of algae in soils even for amendment purposes also brings minerals that will be assimilated as nutrients by the plant. This is why the use of algae in the soil amendment process is often accompanied by a fertilization effort of the soil.

The coralline algae *Lithothamnium calcareum* (*Phymatolithon calcareum*) (Cabioc'h et al. 1992) and *L. corallioides* belonging to the class Rhodophyceae (red algae) have a calcified thallus (see Figure 2.1).

These species can live up to a depth of 30 m and, once dead, their initially red thallus fades to a whitish color (Cabioch 1970; Férec and Chauvin 1987). They constitute a coarse sediment called "maerl", which is collected by dredging.

The use of these calcareous sediments as a soil conditioner is an ancient practice among Celtic populations. In CE 79, Pliny mentioned the use by the Gauls of such an amendment called "marga", "*marne*" in French and "*marl*" or "*maërl*" in Breton (Grall and Hall-Spencer 2003). Brittany concentrates the most significant and the most extended deposits of Europe (Grall and Hall-Spencer 2003). These deposits, which are estimated to be 5,000 years old, have traditionally been exploited for the production of agricultural amendments.

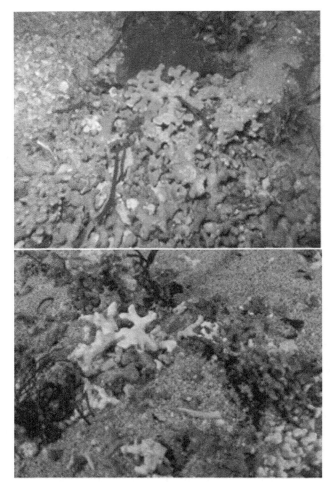

Figure 2.1. *Lithothamnium calcareum (Phymatolithon calcareum) (photo credits © F. André, 2021; E. Amice, CNRS 2021). For a color version of this figure, see www.iste.co.uk/fleurence/algae.zip*

For agricultural use, the maerl after harvesting is crushed into granules for spreading in the soil. The addition of this alga, rich in calcium and magnesium carbonates, allows the pH of acidic soils to be raised before cultivation (see Table 2.1). Its composition in trace elements also gives it a fertilizing activity that can be used in agriculture (see Table 2.2).

Although mainly used by farmers to correct the pH of soils that are too acidic, maerl is also available for home gardeners. Marketed in small packages, this product is generally distributed under the name "Lithothamnion" (see Figure 2.2).

Minerals and mineral salts	Content (in kg)
Calcium carbonate	800–850
Magnesium carbonate	100–150
Lime phosphate	8–10
Phosphorus	6–7
Sulfur	4–7
Silica	4–6
Potassium	1–2
Sodium	1–2

Table 2.1. *Composition of minerals and mineral salts in 1 ton of dead maerl usable in agriculture (from Augris and Berthou 1990)*

Trace elements	Content (in g)
Iron	2,200–2,500
Iodine	500–1,500
Manganese	350–500
Boron	80–350
Pewter	80–120
Fluorine	40–50
Copper	15–50
Zinc	15–50
Titanium	20–40
Molybdenum	3–4
Cobalt	1–2

Table 2.2. *Trace element composition of 1 ton of dead maerl usable in agriculture (from Augris and Berthou 1990)*

Figure 2.2. *Example of a consumer product with maerl as its basis marketed as an amendment and/or fertilizer (photo credit © I. Perez, Masso compagnie 2021)*

2.2. Soil fertilization

The use of seaweed as a soil fertilizer is a safe practice (see Chapter 1). Algae fertilizers are available in solid or liquid extract form (see Table 2.3). In the first case, the solid fertilizer, generated from a mixture of algae and manure, is mechanically applied to the soil (digging, plowing). In the second case, the liquid extract is spread by spraying the soil. Foliar spraying of the extract or soaking of the roots of young shoots in the extract is also possible. In these last two cases, it is not a question of fertilizing the soil but of a direct action on the plant (see Figure 3.1).

Currently, many liquid products, mainly based on brown algae (*Ascophyllum nodosum*, *Laminaria digitata*, *Fucus*), are marketed with a fertilizer claim (see Table 2.3). They mainly act on plant development and growth, thus meeting the expected effects of fertilizing a soil beforehand. However, many studies have suggested that the amount of nutrients provided by the application or spraying of algal extracts is insufficient in accounting for a fertilizing activity. Indeed, the mode of action of algal extracts is most often associated with their phytohormonal activities (auxinic, cytokinin), or even activating the soil's bacterial flora, the latter facilitating the nutritional availability of naturally present mineral elements. This mechanism sometimes leads to the reclassification of these "fertilizers" as products that stimulate plant or biostimulant development (see section 3.1).

Country	Liquid fertilizer	Solid fertilizer	Algal species	Producer
France	Seaweed XTRACT	–	*Ascophyllum nodosum*	Platinium Nutrients
France	–	Algal compost	Not specified	ESAT 4 Vaulx Jardin
France	–	Manure and algae	Not specified	Or Brun
France	–	Old-fashioned fertilizer with manure and algae	Not specified	Tonusol
France	Plantalg algal extract	–	– *Ascophyllum* sp. – *Laminaria* sp. – *Fucus* sp.	Plantin
France	Agrocéan laminactif Agrocéan laminaplus	–	*Laminaria* sp.	Agrocéan
Netherlands	Alga grow	–	Not specified	Plagron
Netherlands	Alga bloom	–	Not specified	Plagron
United Kingdom	SeaFeed Xtra	–	*Ascophyllum nodosum*	Envii
United Kingdom	Shropshire	–	*Ascophyllum nodosum*	Seachem
Norway	–	Kelp	*Ascophyllum nodosum*	Terralba
United States	Maxicrop Seaweed + Tomato fertilizer	–	*Ascophyllum nodosum*	Maxicrop
United States/Italy	–	Maxicrop Seaweed + Tomatoes fertilizer	*Ascophyllum nodosum*	Maxicrop
United States	Maxicrop Premium Seaweed + Fertilizer	–	*Ascophyllum nodosum*	Maxicrop
United States	–	Maxicrop Organic Kelp meal	*Ascophyllum nodosum*	Maxicrop/Valagro
United States	Maxicrop Natural fertilizer	–	*Ascophyllum nodosum*	Maxicrop
United States	Maxicrop Indoor Plant Food	–	*Ascophyllum nodosum*	Maxicrop

Table 2.3. *Examples of some products (composts, powders, liquid extracts) with seaweed as their basis under the main commercial claim of fertilizer (author's note: some products associate a biostimulant effect as a secondary claim)*

Many algal extracts marketed as fertilizers sometimes mention the presence of this biostimulant activity in a secondary commercial claim.

Currently, seaweed-based products and more particularly liquid extracts are mainly used in agriculture for their stimulating properties on plant productions. Agrobiology and its advances in understanding the mechanisms involved in the agronomic performance of crops have led to a different approach to the effect of algae on crops. The latter has allowed the industry to develop a scientific argument that goes beyond the traditional claim of soil fertilization associated with the use of algae in agriculture.

2.3. Improvement of composts for agricultural use

Algae can also be used for soil fertilization by adding them to vegetable or animal compost (animal droppings), which will be used later as fertilizer.

The addition of brown algae (*Fucus serratus*) and green algae (*Ulva lactuca*) increases the temperature of the aerobic fermentation of an agroforestry compost, which can reach 80°C in the core of the compost, thus promoting the fermentative process (Béchu et al. 1988, see Figure 2.3). In the absence of algae, the internal temperature of the compost does not exceed 70°C, regardless of the incubation time (see Figure 2.4).

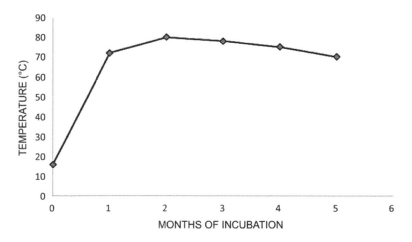

Figure 2.3. *Effect of the addition of algae (*Fucus serratus*) on the internal temperature of an agroforestry compost (from Béchu et al. 1988)*

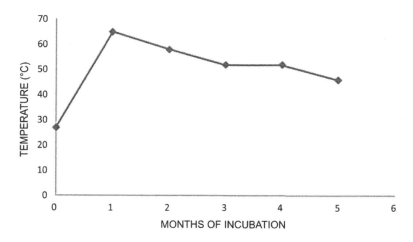

Figure 2.4. *Evolution of the internal temperature of an agroforestry compost without the addition of algae (from Béchu et al. 1988)*

The addition of algae to compost improves the fermentation process by promoting the development of cellulolytic and lignolytic microbial strains. This practice leads to a reduction of half of the degradation time of plant compounds. Agroforestry compost is also characterized by an enrichment of its chemical composition, which gives it a higher agronomic value than conventional agroforestry compost (see Tables 2.4 and 2.5).

	Total nitrogen (%)	Phosphorus (%)	CaO (%)	MgO (%)	Potassium (%)
Poplar bark	0.75	0.26	3.9	0.33	0.61
Poplar bark + *Fucus* sp.	0.88	0.27	7.4	0.61	1.02
Poplar bark + *Ulva* sp.	0.93	0.42	6.2	0.46	0.29
Plane tree bark + seaweed mix*	1.16	0.58	10.2	0.69	0.75

Table 2.4. *Chemical composition of different composts with and without algae (from Béchu et al. 1988). *Ulva lactuca, Fucus serratus, F. vesiculosus, Laminaria digitata, L. hyperborea, L. saccharina*

	Agroforestry compost	Agroforestry compost + seaweed mix*
Magnesium (%)	0.045	0.247
Iron (%)	0.036	0.20
Copper (mg/kg)	1.8	4.7
Manganese (mg/kg)	4.0	4.0
Sulfur (%)	0.03	0.05
Boron (mg/kg)	0.1	0.35
Molybdenum (mg/kg)	3.8	9

Table 2.5. *Trace element composition of a conventional agroforestry compost and a compost incorporating a mixture of algae (from Béchu et al. 1988).* *Ulva lactuca, Fucus serratus, F. vesiculosus, Laminaria digitata, L. hyperborea, L. saccharina

The use of algae as an enhancer of composts for agricultural use is not limited to agroforestry composts. Algae can be mixed with other sources such as straw, various plant wastes or animal manure (cows or horses). Such experiments are often carried out with ulva, which represents an easily available biomass due to its accessible and ubiquitous character since it is present on all coasts. In Paracas Bay, Peru, a site where large quantities of ulva are stranded, algae are added to different types of compost in proportions of 9, 17 and 28% of the total volume of the composted biomass (see Table 2.6, Wosnitza and Barrantes 2005).

Ulva can be added without rinsing, after rinsing or as a powder. All types of compost, including those without the addition of algae, quickly reach the thermophilic phase (> 40°C), which corresponds to the colonization of the compost by microorganisms that will degrade the plant biomass. The addition of 9% ulva in powder form allows the compost to reach a temperature of 62°C and the longest thermophilic phase (15 days). These high temperature conditions and long duration of the thermophilic process guarantee the destruction of pathogens. In addition, the increasing use of this type of compost (C6) leads to improved growth of maize plants (see Figure 2.5).

In most composts, algae are used in small quantities compared to other compounds, and they mainly play a role as activators of aerobic fermentation. However, there are cases where algae are the main constituents of the organic compost.

	Compost 1 (C1)	Compost 2 (C2)	Compost 3 (C3)	Compost 4 (C4)	Compost 5 (C5)	Compost 6 (C6)
Straw/Ipomoea (%)	75	70	62	54	70	70
Cow manure (%)	25	21	21	18	21	21
Ulva sp. (%)	0	9[a]	17[a]	28[a]	9[b]	9[c]

Table 2.6. *Composition of the different composts tested with and without ulva from Paracas Bay (from Wosnitza and Barrantes 2005). a) Washed seaweed. b) Unwashed seaweed. c) Powdered seaweed*

In some countries, such as Pakistan, stranded algae are harvested for algal composts incorporating a reduced proportion of animal dung, principally cow dung (Haq et al. 2011). The algae used in this process (*Ulva fasciata*, *Chondria tenuissima*, *Sargassum* sp.) are very common on the Pakistani coast.

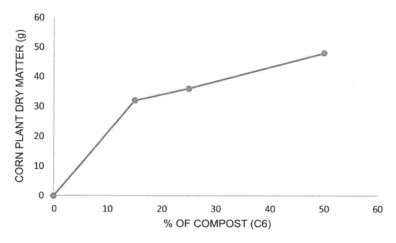

Figure 2.5. *Effect of adding increasing levels of C6 compost containing ulva powder on the growth of corn plants (from Wosnitza and Barrantes 2005)*

The compost consists of two volumes of algae (750 kg wet matter) and one volume of cow dung (250 kg wet matter). The whole is covered with a layer of straw to insulate the compost. After 40 days of maceration, the compost is left to mature for another week. After this operation, it is dried and sieved. This composting process allows the conversion of 70% of the algal biomass into organic compost. The resulting compost has a neutral pH (7.1), a reduced moisture content (4.95%) and a C/N (carbon/nitrogen) ratio of 7:1 compared to 18:1 in the initial algal biomass (see Table 2.7).

The application of such a compost in the seed germination of maize (*Zea mays*) was found to be significantly better than that of animal compost made from cow dung (see Table 2.8).

Composts are organic fertilizers used in traditional agriculture, especially in developing countries. Algae can be integrated in a vegetal or mixed compost (integrating animal droppings). They favor, by their presence, the fermentation within the compost and contribute to improving the efficiency of the fermentative process. They can also represent the only fermentable source of compost. This is the case with algal composts, which were widely used on the North European coasts.

	Algal biomass	Algal compost
pH	6.8	7.1
Humidity (%)	83.9	4.95
Solid mass (%)	16.10	95.05
Inorganic matter (%)	39.70	72.49
Carbon (%)	33.13	15.11
Nitrogen (%)	1.88	2.3
C/N	18:1	7:1
Odor	Strong marine odor (unacceptable)	Classic compost odor (acceptable)

Table 2.7. *Main physico-chemical characteristics of algal compost obtained after mixing with cow dung (from Haq et al. 2011)*

	Animal compost (cow dung)	Algal compost	Commercial organic compost (biogold)
Germination rate of corn seeds (%)	60	78	83

Table 2.8. *Comparative efficacy of different composts on the germination of corn seeds (from Haq et al. 2011)*

Traditionally, algae are used to modify the structure of the soil, correct its pH and bring nutrients that are favorable for plant growth. As such, they are amendments and fertilizers of natural origin, conducive to the development of sustainable agriculture by limiting the contribution of chemical inputs into the soil.

3

Biostimulation Activities on Plant Productions

Seaweed is traditionally used for soil amendment and fertilization of soils (see Chapter 2) before the cultivation of vegetable productions of agronomic interest (vegetable, fruit or cereal crops).

Many commercial products currently available claim to stimulate plant growth, induce defense mechanisms against parasitic infections (viruses, bacteria, fungi) and provide resistance to abiotic stresses (water stress, thermal stress, etc.) (see Table 3.1). These activation properties in the plant's physiology are mainly associated with the use of liquid algal extracts. These products are applied directly to the plant by pulverization, seedling treatment or seed dipping (see Figure 3.1). They can also be spread on the topsoil or introduced into the nutrient solution feeding hydroponic crops. Solid products in the form of granules, flours or even seaweed composts can also be incorporated into the soil to stimulate plant development (Asco-Root, see Table 3.1).

The understanding of the mechanisms involved, although patchy, is progressing steadily. The role of certain molecules in disease resistance or plant growth such as polysaccharides (laminarin) or phytohormones (auxinscytokinins) has often been demonstrated.

These algal extracts with stimulating activities of plant development are qualified as "biostimulants". They are defined as follows:

> Biostimulants are products other than fertilizers that promote plant growth when applied in small amounts (Chatzissavvdis and Therios 2014).

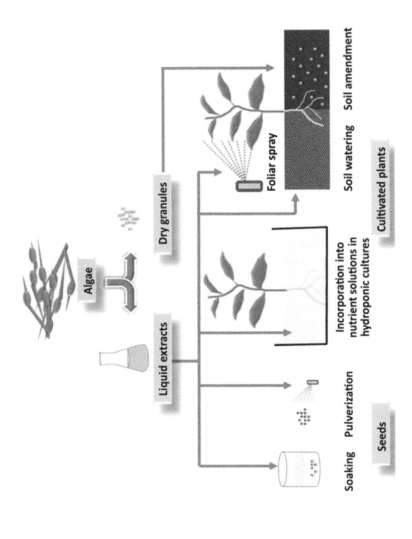

Figure 3.1. *Application modes of liquid or solid extracts of algae as biostimulants of plant crops (source: Y.-F. Pouchus 2021). For a color version of this figure, see www.iste.co.uk/fleurence/algae.zip*

3.1. Stimulation of growth

Liquid extracts or solid algal products (see Table 3.1) are used as stimulants in plant development and growth. The effects induced and the mechanisms involved depend on many parameters including the mode of application. There are two main modes of application, namely the spreading of whole algae, manures, flours or even granules in the soil or the spraying of liquid extracts. The application of solid products is often associated with a fertilizing action of the soil and these products are mainly listed as fertilizers (see section 2.2). In addition to this fertilizing action, manure–algae mixtures or granules are also marketed for their stimulating virtues on the enrooting of plants and thus on plant development. This observation highlights the existence of a fine line between the claims of fertilizers and biostimulants that can sometimes be applied to the use of algae in agriculture.

However, the application of solid products has some disadvantages. Indeed, the direct use of algae, even finely ground, without deep turnover of soil, has an inhibiting effect and therefore does not stimulate plant growth. This is due to the production of algal sulfur compounds that induce a delay in the germination of seeds and in the growth of plants. Without mechanical turning of the topsoil, the inhibition of germination and plant growth is not lifted until after 15 weeks (Craigie 2011). The use of composts, whether made up solely of algae or some other mixture, does not lead to this kind of drawback.

Applications of liquid extracts to stimulate plant growth and development are used as foliar sprays or on seeds. They can also be applied by impregnation of the soil or by addition to the nutrient solutions used in hydroponics (see Figure 3.1).

The spraying of liquid extracts at the foliar or soil level allows a more rapid assimilation of algal compounds by the aerial part of the plant or by the surrounding rhizosphere (see Figure 3.1). Algal liquid extracts, usually composed of brown algae (e.g. *Ascophyllum nodosum*), are rich in fucan-like algal polysaccharides (see Figure 3.2) which, in their rhizobacteria-stimulating degraded form, are involved in the plant's nutrition and development process (Craigie 2011; Fleurence 2022).

Many products marketed as plant growth biostimulants do not target a specific species. However, there is a range of products that are more specifically targeted to a crop (see Table 3.2). These products are made from algae of the genera *Ecklonia*, *Ascophyllum*, *Sargassum* and *Cystoseira*. Some of them are also made from red algae of the genus *Kappaphycus*. The crops targeted by these products can be ornamental such as geranium, fruit crops such as citrus or vine or even vegetable crops such as tomatoes or cucumbers. The application methods of these liquid products are identical to those previously described (see Figure 3.1).

Country	Company	Product	Algal species
South Africa	Kelp Products Ltd	Kelpak[1]	*Ecklonia maxima*
Germany	Neomed Pharma GmbH	Algifol[1]	*Ascophyllum nodosum*
Australia	Seasol International Ltd	Seasol[1]	*Durvillaea potatorum*
Australia	Fair Dinkum Fertilizers	Seaweed[1]	*Durvillaea antarctica*
Canada	OrganicOcean	Stimulagro[1]	*Ascophyllum nodosum*
Canada	OrganicOcean	Asco-Root[2]	*Ascophyllum nodosum*
Canada	Acadian Seaplant Ltd	Kelp Meal[3]	*Ascophyllum nodosum*
China	China Ocean University Project Development Ltd.	Sea winner[1]	Not specified
China	Leili	Algreen[1]	Not specified
France	Setalg	Algovert[1]	– *Laminaria digitata* – *Ascophyllum nodosum*
France	Goëmar Laboratories	Goëmar[1]	*Ascophyllum nodosum*
India	PI Industrie Ltd.	Biovita[1]	*Ascophyllum nodosum*
Ireland	Bioatlantis Ltd.	Ecolicitor[1]	*Ascophyllum nodosum*
United Kingdom	Micromix Plant Health Ltd.	GS35[1]	*Ascophyllum nodosum*
United States	Atlantic Laboratories Inc.	Seacrop[1]	*Ascophyllum nodosum*
United States	Green Air Products Inc.	Bio-Genesis[1]	*Ascophyllum nodosum*

1. Liquid product. 2. Product in granular form. 3. Product in flour form.

Table 3.1. *Examples of some products marketed for their properties in stimulating plant growth or inducing resistance mechanisms to biotic and abiotic stresses (from Chatzissavvidis and Therios 2014 and Sharma et al. 2014)*

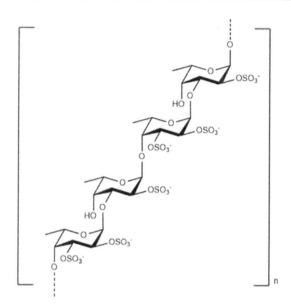

Figure 3.2. *Typical structure of a fucoidan from the brown alga* Ascophyllum nodosum *(Fleurence 2018)*

The application of algal extracts to plants results in the establishment of different mechanisms influencing plant growth and development. The administration of algal extracts, especially of brown algae, containing degraded polysaccharides improves the solubility of certain elements present in the soil. This is the case for copper, cobalt, manganese and iron, which are important trace elements for plant development (Fleurence 2022).

Liquid products based on brown algae (e.g. *A. nodosum*) also have an activating effect on the bacterial flora of the soil during their application. This effect is mainly due to the presence of fucoidan or alginate (see Figure 3.3) which will contribute to the aeration of the topsoil and the stimulation of rhizobacteria (Fleurence 2022). These bacteria, known by the acronym PGPR (plant growth-promoting rhizobacteria), are part of the root environment of the plant. They contribute to the improvement of atmospheric nitrogen fixation and to the solubilization of phosphate contained in the soil. The activation of the PGPR bacterial flora by algal extracts is therefore considered to be an alternative to the use of nitrogen fertilizer. As such, such a practice from agrobiology is part of a new agricultural approach that can be qualified as ecoresponsible.

Product	Algal species	Effect	Culture
Kelpak	*Ecklonia maxima*	Increase in the mass of cuttings	Geraniums
Experimental product	*Ascophyllum nodosum*	Increase in stem, leaf and bulb mass	Lilies
Stimplex	*Ascophyllum nodosum*	Stimulation of shrub growth	Lemon trees
Kelpak	*Ecklonia maxima*	50% increase in grain yield	Barley
Exp	*Kappaphycus alvarezii*	Increased crop yield	Wheat
Algamino/Goëmar	*Sargassum* sp.	15–25% increase in seedling fresh matter	Corn
Kelpak	*Ecklonia maxima*	Increased plant growth	Swiss chard
Exp	*Cystoseira barbata*	Increased seed germination at 15°C and 25°C	Eggplant
Algal 30	*Ascophyllum nodosum*	Increase in yield	Sweet pepper
Exp	*Ascophyllum nodosum*	Increase in yield	Potatoes
Goëmar	*Ascophyllum nodosum*	Increased vegetative growth	Apple trees

Table 3.2. *Example of liquid products with a base of seaweed extracts used for the stimulation of the growth or development of specific crops*

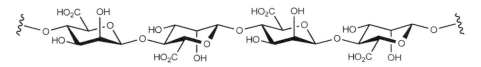

Figure 3.3. *Example of an alginate molecule (polymer of mannuronic acids or poly M) (Fleurence 2018)*

Induction of rhizosphere bacteria is not limited to the input of liquid macroalgal extracts into the soil. The addition of microalgae, or even cyanobacteria, contained in sludge from mudflats in particular (see Figure 3.4) can also be considered (Zhang et al. 2014). This type of experimentation was carried out using sludge re-floated from Lake Tai (or Taihu), which is a large Chinese lake with a surface area of 2,250 km² straddling the provinces of Jiangsu and Zhejiang[1].

Figure 3.4. Examples of mudflat sludge rich in microalgae and cyanobacteria used for their biostimulatory properties for plant growth (photo credits © A. Barnett, P. Rosa, 2021). For a color version of this figure, see www.iste.co.uk/fleurence/algae.zip

A compost developed from cattle manure and microalgal sludge was tested on the development of a bacterial strain SQR 69 belonging to the PGPR family (Zhang et al. 2014). This compost was used as a fermentative medium for the bacterial culture of SQR 69.

A compost consisting only of cow manure was used as a solid fermentative control medium. In the presence of algal sludge and after six days of fermentation, the cell density of SQR 69 rhizobacteria was 5.83×10^7 cells/g compared to 0.93×10^7 cells/g with the control compost (see Figure 3.5).

This experiment suggests that sludge consisting of microalgae and probably cyanobacteria has an inducing effect on the growth of soil rhizobacteria, which play a fundamental role in plant nutrition and growth.

Other experiments have shown the effectiveness of microalgae or cyanobacteria in stimulating plant growth in many crops (Kang et al. 2021). This is notably the case for cucumber cultivation where the presence of the alga *Chlorella vulgaris* has a positive effect on root growth, and thus on plant development (see Table 3.3). Plant production of onion is also improved when a mixture of cow dung and cyanobacteria belonging to the species *Spirulina platensis* (*Arthrospira platensis*) is incorporated into the crop (Kang et al. 2021). In tomatoes, the addition of the microalga *C. vulgaris* also improves root development and shoot growth.

---

[1] https://en.wikipedia.org/wiki/Lake_Tai.

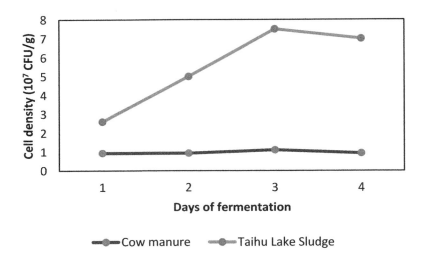

Figure 3.5. *Effect of Lake Taihu microalgal sludge and cow manure on the growth of rhizobacterial strain SQR 69 (from Zhang et al. 2014). For a color version of this figure, see www.iste.co.uk/fleurence/algae.zip*

The use of mixtures composed of microalgae or cyanobacteria and rhizobacteria has also shown the synergistic effect of these microorganisms in terms of biostimulation of plant growth.

The mixed application of *C. vulgaris* and the rhizobacterium *Azotobacter* sp., which fixes atmospheric nitrogen when growing lettuce, increases the plant mass of the crop very significantly under these conditions (see Table 3.3). A similar result is obtained when this microalga is supplied in association with rhizobacteria of the genera *Bacillus* or *Azospirillum*.

In onion, a positive synergistic effect on growth is also observed when the cyanobacterium *S. platensis is* combined with the rhizobacterium *Pseudomonas stutzeri*.

The action of algae, whether macro- or microalgae, on the bacteria of the rhizosphere is one of the mechanisms involved in the biostimulation of plant growth and development. The liquid extracts are mainly elaborated from macroalgae and more particularly from brown algae (e.g. *Ascophyllum* sp., *Laminaria* sp.). Microalgae and cyanobacteria are often introduced via solid residues such as sludge, whether of natural origin such as mudflats (see Figure 3.4) or artificial origin such as sewage sludge. These products with their stimulating properties for the bacterial flora of the rhizosphere are called biofertilizers and are fully in line with organic agriculture.

Plant production	Microalgae/cyanobacteria*	Bacteria	Effect
Tomato	Chlorella vulgaris	–	Stimulation of shoot and root growth
Tomato	Chlorella vulgaris	– Azotobacter sp. – Bacillus licheniformis – Bacillus megaterium	Increase in plant size
Onion	Spirulina platensis*	Pseudomonas stutzeri	Increased plant growth
Lettuce	Scenedesmus quadricauda	–	Increased plant growth and leaf protein content
Cucumber	Chlorella vulgaris	–	Induction of root growth
Bean	Anebaena cylindrica*	– Azospirillum brasilense – Rhizobium sp.	Stimulation of plant growth and increase in grain production (+84%)
Corn	Anaebaena cylindrica	Azospirillum brasilense	Increased production output

Table 3.3. *Biostimulatory effect of microalgae or cyanobacteria* used alone or in synergy with rhizobacteria (PGPR) on growth of plants of agronomic interest (from Kang et al. 2021)*

The stimulation of growth by algal extracts also relies on different mechanisms from those linked to the activation of the rhizobacterial flora.

Many liquid products obtained from brown algae have been found to contain growth hormones such as cytokinins or auxin (Fleurence 2022). Cytokinins are known for their cell division activating properties and promote the opening of leaves, thus contributing to the development of the plant. Auxin or indole 3-acetic acid (see Figure 3.7) acts on the increase of cell size and stimulates mitosis. It also contributes to the elongation of stems and branches and to the neoformation of organs (Kofler 1963).

Auxin strongly stimulates cell growth when associated with cytokinins and thus participates in plant development.

The presence of cytokinins in the algal extract obtained via alkaline hydrolysis of the brown alga *Durvillaea potatorum* was formally identified in the mid-1980s (Tay et al. 1985). This extract contains, in particular, trans-zeatin, a cytokinin hormone originally isolated from maize, as well as its ribosyl derivative (see Figure 3.6).

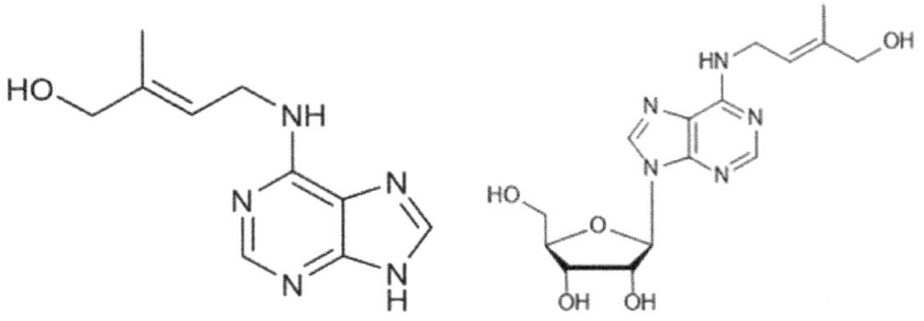

Figure 3.6. *Molecular structures of trans-zeatin (a) and trans-zeatin-9-β-D-ribofuranoside (b) (Fleurence 2022)*

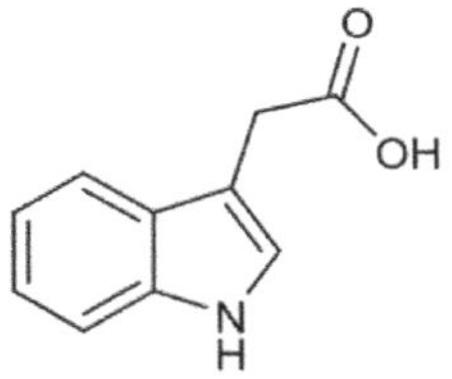

Figure 3.7. *Indole 3-acetic acid (auxin)*

Many brown algae belonging to the order Fucales or Laminariales are also listed as producers of growth hormones such as cytokinins, auxin or gibberellins (see Table 3.4 and Figure 3.7).

In addition to the presence of hormones that stimulate plant growth and development, some brown algae also produce abscisic acid (see Figure 3.8) which is a germination inhibiting phytohormone, and thus a regulatory factor for plant development.

Phytohormones	Types of algae
Auxin and derivatives	– *Ascophyllum, Fucus* (Fucales) – *Laminaria, Undaria, Macrocystis* (Laminariales)
Cytokinins	– *Ascophyllum, Cystoseira, Fucus, Sargassum* (Fucales) – *Ecklonia, Macrocystis* (Laminariales)
Gibberellins	– *Cystoseira, Fucus, Sargassum (Fucales)* – *Ecklonia* (Laminariales) – *Petalonia* (Ectocarpales)
Abscisic acid	– *Ascophyllum* (Fucales) – *Laminaria* (Laminariales)

Table 3.4. *Families of phytohormones produced by different genera of brown algae belonging to the orders Fucales, Laminariales and Ectocarpales (from Craigie 2011 and Chatzissavvidis and Therios 2014)*

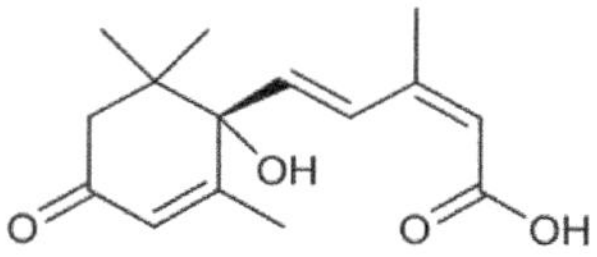

Figure 3.8. *Abscisic acid (Fleurence 2022)*

The presence of cytokinins, auxins or related activities is described in many species of brown algae and on commercial products derived from them.

Indole 3-acetic acid (IAA) or auxin (see Figure 3.7) is formally characterized in the product Maxicrop, which is formulated from *A. nodosum* extract (Sanderson et al. 1987). The IAA content is estimated to be 6.63 µg per gram of dry product.

A wide variety of products obtained from very different processes (see Table 3.5) were found to contain phytohormones or molecules with related hormonal activities.

The maintenance of phytohormonal activity in these products suggests that the extraction processes used do not alter the biological activity of the extracts. The presence of zeatin, its ribosyl derivative or the hormone isopentenyladenine is reported for the products Seamac, Kelpak, Marinure, Maxicrop and SM3 (see Table 3.5). These products activate callogenesis in the soybean model and significantly improve root number in mung beans (see Figure 3.9).

Species of algae	Commercial product	Extraction process	Hormone
Laminariales and *Fucales*	SM3	Aqueous extraction	– Zeatin – Ribosylzeatin
Ascophyllum nodosum	Maxicrop	Alkaline hydrolysis	Ribosylzeatin
Ascophyllum nodosum	Marinure	Alkaline hydrolysis	Ribosylzeatin
Ascophyllum nodosum	Seamac	Alkaline hydrolysis	Isopentenyladenine
Durvillaea potatorum	Seasol	Alkaline hydrolysis	– Zeatin – Ribosylzeatin
Ecklonia maxima	Kelpak	Cellular breakdown	– Zeatin – Ribosylzeatin

Table 3.5. *Examples of commercial products in which the presence of phytohormones is described (from Tay et al. 1985 and Stirk and van Staden 1997)*

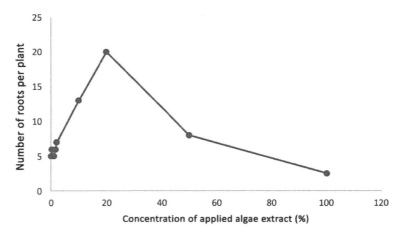

Figure 3.9. *Effect of the commercial product Seamac on the increase in the number of roots in mung beans as a function of the concentration of applied product (from Stirk and van Staden 1997)*

However, the concentration of phytohormones in algal extracts is considered insufficient by some authors to explain biostimulation effects on plant growth and development (Wally et al. 2013). Experiments performed on the model plant *Arabidopsis thaliana* suggests a somewhat different mechanism of action than that commonly considered, which is that of direct action of algal phytohormones on crop development. The application of commercial products with *A. nodosum* on *A. thaliana* as their basis shows in particular that the administration of an extract containing a cytokinin concentration of 3 mg.L^{-1} results in the presence within the plant of a hormonal activity corresponding to a cytokinin concentration of 5 mg.L^{-1} (Kahn et al. 2011; Fleurence 2022).

Application of a commercial extract of *A. nodosum* to this model plant indeed induces an endogenous production of cytokinins and more particularly of trans- and cis-zeatins. A 50% increase in cytokinin concentration in leaves is reported after spraying the algal extract (Wally et al. 2013; Fleurence 2022). The increase in trans-zeatin production is notably significant 24 hours after the application of the extract. A similar effect on cis-zeatin production (see Figure 3.10) is observed, this time after 114 hours. An overexpression of the genes involved in the production of these two types of cytokinins is also recorded during this period.

In this case, the algal extract appears more as a stimulant of the endogenous pathway of phytohormone biosynthesis than as an exogenous supplier of growth hormones.

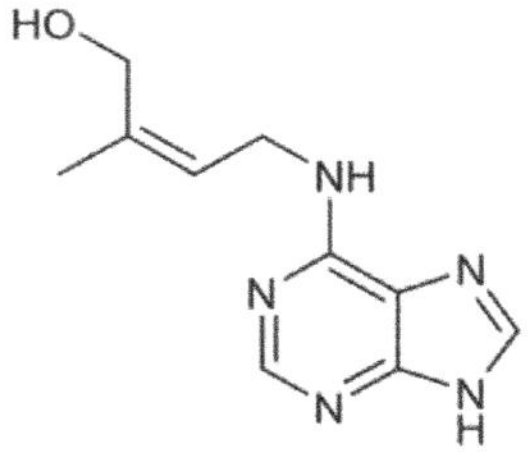

Figure 3.10. *Cis-zeatin*

The mechanisms induced by algal extracts to stimulate plant growth and development are often associated with an improvement of the plant nutrition process or with the direct or indirect action of phytohormones. However, the activity of many algal extracts or commercial products is evaluated on the basis of observed agronomic effects without specific research on the mechanisms involved.

This is the case of green or brown algal extracts tested on the growth of vegetable crops such as tomatoes (*Solanum lycopersicum* L.). Extracts obtained from *Ulva lactuca* (green alga), *Caulerpa sertularioides* (green alga), *Padina gymnospora* (brown alga) or *Sargassum liebmannii* (brown alga) collected from the Mexican coast appear to be biostimulants of germination and growth of tomato plants grown in greenhouses (Hernandez-Herrera et al. 2014). The best effects on seed germination are observed with extracts of *U. lactuca* and *P. gymnospora* at the 0.2% concentration (see Table 3.6). In both species, an antagonistic effect on tomato seed germination is noted when the extract concentration is increased. Moreover, extracts from *Caulerpa* and *Sargassum* were less effective in inducing seed germination. These results highlight, on this type of culture, the influence of the nature of the alga and the concentrations used on the germination of tomato seeds.

Extract	Germination (%)	Germination time (days)
Control	31	5.9
Ulva lactuca (0.2%)	75	5.6
Ulva lactuca (0.4%)	55	5.8
Ulva lactuca (1%)	50	5.8
Padina gymnospora (0.2%)	76	5.6
Padina gymnospora (0.4%)	66	5.7
Padina gymnospora (1%)	46	5.8
Caulerpa sertularioides (0.2%)	1	6.3
Caulerpa sertularioides (0.4%)	12	6.2
Caulerpa sertularioides (1%)	5	6.5
Sargassum liebmannii (0.2%)	1	5.9
Sargassum liebmannii (0.4%)	2	5.9
Sargassum liebmannii (1%)	1	6.0

Table 3.6. *Effect of green or brown seaweed extracts on germination of tomato seeds (from Hernandez-Herrera et al. 2014)*

The effect of algal extracts on the development of the tomato plant after germination is also well documented. It is evaluated by measuring several parameters (length of the first leaf, initial root, dry weight). The best development of the first leaf (plumule) is observed after the application of *U. lactuca* and *P. gymnospora* extracts at the concentration of 1% (Hernandez-Herrera et al. 2014). Compared to control seeds, these extracts induce in treated germinated seeds a significantly increased plumule development by a value of 62.2% with *U. lactuca* and 84.4% with *P. gymnospora*.

In contrast, at this concentration, extracts of *C. sertularioides* and *S. liebmannii* were less effective for plumule development. There is therefore a dose effect of the application of algal extracts on plumule development, this effect varying according to species.

A similar observation was made concerning the development of the initial root or radicle. The length of the latter is higher (6 cm against 4 cm for the control) after treatment with the extract of *P. gymnospora* at the concentration of 0.2%. However, it decreased with increasing extract concentration. More surprisingly, the application of the *Caulerpa*-based extract limits the length of the radicle suggesting an inhibiting effect (−25%) on its development.

Regarding the development of plant biomass, the highest dry mass is mainly observed at the 1% concentration for *P. gymnospora* and *U. lactuca* extracts.

In addition to these effects related to algal species and concentrations employed, the mode of administration also appears to have an influence on the development of tomato plants in greenhouses.

Application via soil drench with extracts appears to be more effective than pulverization in terms of plant elongation (+6.7%).

These results highlight the multiplicity of factors of variation associated with the use of algal extracts (species used, dose effect, mode of application, phase of development of the plant). They support the need to develop standard procedures, particularly in terms of application methods, regardless of the crop production concerned.

The resource used for extract production also depends on the algal diversity present on the coast. For example, in the Indian state of West Bengal, algal extracts are often made from red algae belonging to the genera *Kappaphycus* and *Gracilaria* (Pramanick et al. 2014). The application of such extracts by foliar spray at the concentration of 15% (v/v) improves the grain yield when growing rice very significantly. The latter is increased by 41.5% after the application of *Kappaphycus* extracts and by 35% after administration of *Gracilaria* extracts.

Algal extracts, whether from brown, green or red algae, have a biostimulant effect on the growth of many plants. These extracts, often obtained in small quantities or in a traditional way, do not benefit like commercial products from industrial standardization approaches. The commercial products, standardized from the formulation point of view, are also evaluated on their agronomic impact.

This is the case for Goëmar GA 14, which, when applied by foliar spray, improves the fresh mass of maize shoots by 15–25% (Jeannin et al. 1991). This improvement is notably reflected in an increase in the number of roots and in the mass of the stem.

This product, formulated from the brown alga *A. nodosum*, has also been shown to be efficient in spinach (*Spinacia oleracea*) cultivation (Cassan et al. 1992). When applied as a foliar application at a concentration of 2.5 g.L^{-1}, it improved the fresh matter of spinach leaves produced after eight weeks of cultivation by 12–15%.

Another product, Stimplex, also prepared from *A. nodosum*, has been found to be particularly effective in stimulating the growth of tomato or pepper. A dilute application (0.5% of the extract) improves various growth parameters such as plant length (+40%), plant biomass (+52%) and root length (+59%) (Ali et al. 2019).

In beet (*Beta vulgaris*), the commercial product Kelpak 66 (diluted at 1:500) significantly improved the mass of the plant (+111%) at both the foliar level (+116%) and the root level (+100%), compared to the control plant that was not treated with Kelpak 66 (Featonby-Smith and van Staden 1983).

Algal extracts or commercial products derived from them have proven to be effective growth and development stimulants for a wide range of crops. The main effects are on the development of the root system and the plant biomass. The mechanisms involved are related to the activation of the rhizobacterial flora, to the improvement of the availability of nutrients contained in the soil and to the contribution of phytohormones, whether of exogenous or endogenous origin.

However, the impact of algal extracts is not limited to plant growth. The use of algae as biostimulants also concerns other aspects such as tolerance, or even resistance, to abiotic stress.

The stresses affecting the crops are frequently the hydric stress generating the dehydration of the plant, thermal stress or saline stress.

3.2. Tolerance to water stress

Drought inducing plant dehydration is the most common abiotic stress suffered by crops. It is responsible for a drastic decrease in the yield of plant productions.

Ascophyllum nodosum-based algal extracts are often cited as stimulating tolerance or even resistance to dehydration. In tomato, the effects of administering *A. nodosum* extracts on this type of stress are particularly well documented (Gôni et al. 2018). The extracts tested are commercial products obtained after a classical extraction process at high temperature but performed either at neutral or alkaline pH. Both types of extracts are administrated by foliar spraying at a dilution of 0.33% (v/v) before the application of the dehydrating operation.

The impact of the dehydration process on plant physiology is determined by measuring the relative water content (RWC) of leaves and plant biomass (see Tables 3.7 and 3.8).

After seven days of water stress, the water content of leaves from plants not treated with the algal extract dropped by 14.4% compared to leaves from control, unstressed plants (see Table 3.7).

Preventive treatment with neutral algal extract reduces this water loss to 4.6%.

Water stress application time	RWC Unstressed plants	RWC Stressed plants without ANE 1 input	RWC Stressed plants with ANE 1 contribution
T0	76.3	76.4	76.5
T7	76.6	65.5	73.05

Table 3.7. *Impact of* Ascophyllum nodosum *neutral extract (ANE 1) on relative leaf water content (RWC) of tomato plants subjected to water stress (T0: before stress application; T7: 7 days after application) (from Gôni et al. 2018)*

After seven days, water stress also generates a decrease in the fresh mass of the plant. This decrease is estimated at nearly 47% (see Table 3.8). The preventive use of seaweed extracts significantly reduces this loss. It falls to 32.7% after the application of the neutral extract of *A. nodosum* and to 31.8% after treatment with the alkaline extract. The neutral extract is characterized by high contents of sulfate (16%), fucose (15%) and uronic acids (13%). This composition suggests the majority presence of fucans and alginates in the active extract. Laminarin, a reserve polysaccharide, is also present at 2.3% (w/w).

On the contrary, the contents of these constituents are much lower for the alkaline extract (<10% for each constituent), and the absence of laminarin is reported for this extract.

Treatment	Fresh mass of the plant (g)
Absence of stress	29.71
Water stress without ANE application	15.77
Water stress with ANE 1 application	19.98
Water stress with ANE 2 application	20.27

Table 3.8. *Impact of neutral (ANE 1) and alkaline (ANE 2) extracts of* Ascophyllum nodosum *on the fresh biomass of tomato plants subjected to seven days of water stress (from Gôni et al. 2018)*

The presence of a polyphenol content twice as high for the alkaline extract also distinguishes it from the neutral extract.

The impact of administering an *A. nodosum* extract, by foliar spray or soil impregnation, on vine tolerance to water stress is also well documented (Frioni et al. 2021). The product tested is of manufactured origin (Acadian Marine Plant Extract Powder) and is obtained via alkaline extraction from the alga *A. nodosum*. A dose of 3 g of product per plant is applied whatever the selected mode of administration. In this type of experiment, the vine is subjected to a one-month irrigation stop, thus generating hydric stress, before being watered again. The extract applied by foliar spraying (ANEfl) significantly improves gas exchange at the leaf level and the efficiency of water use by the plant. An improvement in water efficiency of 35% was registered for the treated plant compared to the control plant subjected to identical water stress conditions.

Foliar algal treatment also decreases the physiological consequences of water stress on the vine, better preparing it for the rehydration phase. This results in improved photosynthesis (+ 2.7 µmol $CO_2 m^2 s^{-1}$) and increased photochemical efficiency (Fv/Fm + 0.19).

Other parameters such as leaf mass (+8%) or the amount of soluble sugars in the leaf (+27.3 $mg.g^{-1}$) were significantly increased.

In contrast, the application of the extract by soil impregnation (ANEsoil) induces few changes at the physiological level, either during the period of water stress or during the rehydration process.

As in many cases, the mode of application of the product seems to be a determining factor in the effects induced on the plant.

The mechanisms that could explain the observed effects of algal extracts on the improvement of plant tolerance to drought are partially known. The induction of root growth by these extracts, already implicated in the stimulation of plant development, is one of the mechanisms put forward.

In lodgepole pine (*Pinus contorta*), the application of *A. nodosum* extract to the root system, when seedlings are grown in containers, promotes root development in the spring when transplanted to the forest (MacDonald et al. 2012; see Table 3.9). According to these authors, this increased root development would prepare these plantings to withstand periods of drought that are typically fatal in the first year. The significant effect on the root development of this variety of pine is noted for a dilution of the extract at 1/500. For the other concentrations, a lesser and not very significant effect was observed with respect to the control.

	Total number of roots	Number of long roots (> 5 cm)	Number of short roots (< 5 cm)
Control	140	40	100
ANE (1/750)	160	50	110
ANE (1/500)	190	60	130
ANE (1/250)	130	40	90

Table 3.9. *Effect of the application of* Ascophyllum nodosum *extract at several concentrations on the root development of a pine species (from MacDonald et al. 2012)*

The effect of seaweed extracts on the protection of the root system of plants subjected to drought is also known for a plant of major agronomic interest such as wheat (*Triticum sativum*). Pre-treatment of wheat seeds with *Sargassum* (1.5%, v/v) or *Ulva* (1%, v/v) extracts limits the impact of dehydration phases on root length and mass, leaf area and photosynthetic activity of stressed plants (see Table 3.10; Kasim et al. 2015). In this experiment, the algal extracts are applied by dipping the seeds. This mode of administration is therefore very different from foliar spraying or soil soaking. The seedlings resulting from germination are subjected to two water stresses, SH_1 and SH_2, corresponding to a water deficit of 60–80% of the total water potential. The dehydrating treatments induce a very significant reduction in root mass. It is 51% with the SH_1 treatment and 55% with the SH_2 treatment. After the application of *Sargassum* extract, this decrease is reduced to 34% for the SH_1 treatment and to 38% after the SH_2 treatment. On the contrary, *Ulva* extract is more effective than *Sargassum* extract in protecting the root system from drought. Indeed, in the case of water stress SH_1, *Ulva* extract allows the maintenance of root mass (100%) while preserving most of the root length, which is not the case with *Sargassum* extract. However, the concomitant application of the two extracts generates negative antagonistic effects, especially after the application of the most severe water stress (SH_2), both on the root system and the leaf surface.

Preservation of photosynthetic activity is also a physiological mechanism for drought tolerance. In wheat, both water stresses (SH_1, SH_2) induce a significant decrease in photosynthesis of 9 and 15% (see Figure 3.11). The pre-treatment with *Ulva* extract limits this decrease in photosynthetic activity to 3%. As with the previous parameters, the joint use of *Ulva* and *Sargassum* extracts has a negative effect on photosynthesis and is therefore not protective in the context of a drought.

	Root length (cm)	Root mass (g/plant)	Leaf surface (cm^2/leaf)
Witness	26.9	0.29	5.4
SH$_1$	22.1	0.14	3.4
SH$_2$	16.6	0.13	1.8
SH$_1$ + *Sargassum* sp.	22.7	0.19	5.2
SH$_2$ + *Sargassum* sp.	22.9	0.18	3.8
SH$_1$ + *Ulva* sp.	25.4	0.29	4.9
SH$_2$ + *Ulva* sp.	22.1	0.19	2.8
SH$_1$ + *Sargassum* sp. + *Ulva* sp.	22.4	0.13	2.6
SH$_2$ + *Sargassum* sp. + *Ulva* sp.	18.3	0.09	1.5

Table 3.10. *Effect of dipping application of* Sargassum *sp. and* Ulva *sp. extracts on root development and leaf area of wheat plants under water stress (SH$_1$: 60% water deficit; SH$_2$: 80% water deficit) (from Kasim et al. 2015)*

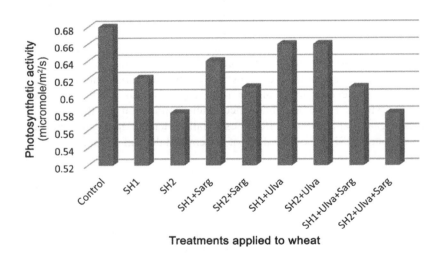

Figure 3.11. *Effect of* Sargassum *sp. (Sarg) or* Ulva *sp. extracts on photosynthetic activity of wheat under water stress (SH1: 60% water deficit; SH2: 80% water deficit) (from Kasim et al. 2015)*

Tolerance, or even resistance, to drought also results from the induction of certain physiological mechanisms limiting water loss or maintaining leaf turgescence. The effects of applying a commercial extract of *A. nodosum* (Algea Valagro) on the drought tolerance of the model species *A. thaliana* are well established (Santaniello et al. 2017). In plants treated with *A. nodosum* extract, a partial closure of stomata is observed from the first day of applying water stress (time 0). This closure is illustrated by a significant decrease in stomatal conductance (55%) compared to that registered in untreated plants (see Figure 3.12).

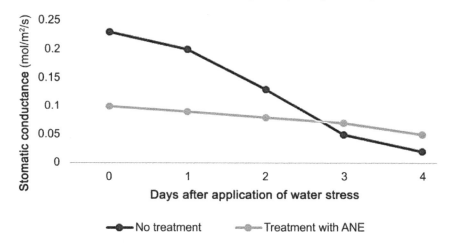

Figure 3.12. *Effect of the application of* Ascophyllum nodosum *extract (ANE) on the stomatal conductance of* Arabidopsis thaliana *leaves after the application of water stress (from Santaniello et al. 2017)*

Similarly, water loss by transpiration is strongly limited from the first day (time 0) of application of water stress in plants treated with the algal extract. The latter decreases by 53% compared to that observed for untreated plants (see Figure 3.13).

The closure of the stomata leading to a limitation of water loss by transpiration thus appears to be a protective mechanism of the plant with regard to drought, the latter being activated early by the application of the algal extract.

The molecular mechanism of action of the extract is also partly known. It acts on the synthesis of abscisic acid, a hormone that negatively regulates the expression of the gene involved in stomatal opening (MYB60).

These physiological and molecular mechanisms significantly improve the water content and survival rate of the water-stressed model plant (see Table 3.11).

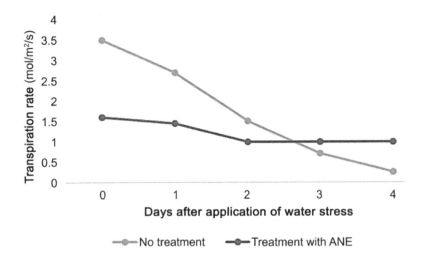

Figure 3.13. *Effect of* Ascophyllum nodosum *extract (ANE) application on the transpiration rate of* Arabidopsis thaliana *leaves after the application of water stress (from Santaniello et al. 2017). For a color version of this figure, see www.iste.co.uk/ fleurence/algae.zip*

	Day 0	Day 1	Day 2	Day 3	Day 4
Treatment-free survival rate (%)	100	100	100	60	10
Survival rate with ANE treatment (%)	100	100	100	100	78
Water content without treatment (g)	300	200	100	50	20
Water content with ANE treatment (g)	300	250	200	150	120

Table 3.11. *Effects of the application of* Ascophyllum nodosum *extract on the survival rate of* Arabidopsis thaliana *and on its water content during the application of water stress (from Santaniello et al. 2017)*

Independently of these mechanisms highlighted on the model plant *A. thaliana*, some authors (Sharma et al. 2014; Fleurence 2022) have mentioned other processes to explain the action of algal extracts on the establishment of plant endurance against drought. These include the involvement of betaines contained in algal extracts (see Figure 3.14). The molecules concerned are aminobutyric acid betaine, aminovaleric acid betaine and glycine betaine (Sharma et al. 2014). Betaines, cytoplasmic osmolytes, play a protective role against osmotic shock generated by water stress following dryness. In particular, they would contribute to better water retention in plant cells subjected to a dehydration process.

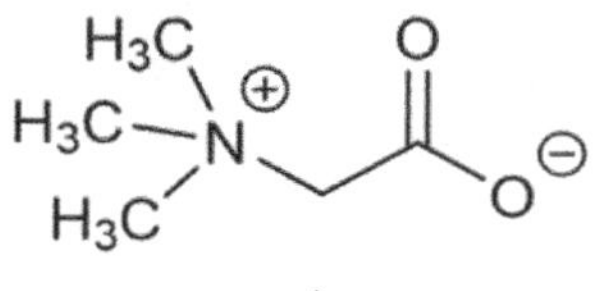

Figure 3.14. *Structures of betaines contained in algal extracts: (a) glycine betaine and (b) aminovaleric acid betaine*

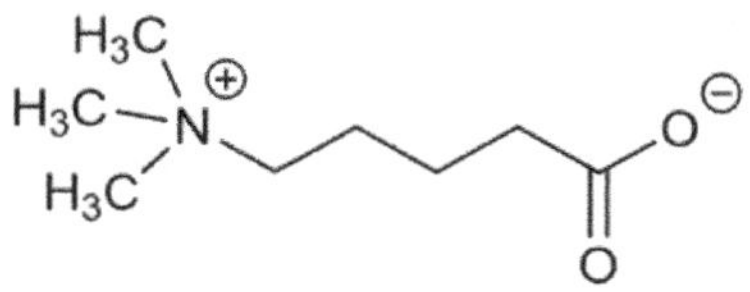

These molecules are characterized in commercially available alkaline extracts of *A. nodosum* (Algifert 25; Blunden et al. 1996). The contents of aminobutyric, aminovaleric and glycine betaines in this type of extract are respectively 120 mg.L^{-1}, 53.2 mg.L^{-1} and 42.6 mg.L^{-1} (Blunden et al. 1996). The betaine content is, however, subjected to a strong seasonal variation (+/–19%) as reported by some authors (MacKinnon et al. 2010).

Algal extracts, and more particularly those obtained from the brown alga *A. nodosum*, appear to be effective in improving tolerance, or even resistance, to drought in certain crops. The protective mechanisms concern the preservation of the root system and the limitation of water losses via an activation of the early closure of stomata. Some molecules, such as the betaines contained in the extracts, also act in the protection process by acting as osmoprotectors. They are also known to activate chlorophyll synthesis in many plants of agronomic interest (barley, wheat, corn, tomato; Blunden et al. 1996). This last property could also explain the protective role of betaines in the face of water stress.

3.3. Tolerance to salt stress

Soil salinization, observed in some tropical or Mediterranean regions, is a limiting factor for agricultural productivity. Some plants called halophytes have a natural tolerance to salts and more particularly to NaCl. Unfortunately, there are very few plants of agronomic interest among the halophytes. The plants primarily valued in cultivation belong to the glycophyte group, whose growth is greatly diminished by the presence of salt in the soil (Levigneron et al. 1995). Salinity mainly affects seed or tuber germination, as is the case for potatoes. Growth and fruiting are also affected by increased saline concentration in the topsoil.

The effects of an *A. nodosum* extract on the protection of 2-year-old avocado (*Persea americana*) subjected to salt stress have been studied in great detail (Bonomelli et al. 2018; Fleurence 2022). In this experiment, plants were subjected to salt stress via the addition of NaCl to the irrigation water. The salt concentration supplied was 9 mM NaCl and treated plants received 1.5 and 2.25 mL volumes of algal extract respectively. Irrigation generating salt stress in the presence or absence of *A. nodosum* extract was administered every two weeks over an eight-month period. The application of salt stress, although moderate (9 mM NaCl), had a strong impact on the development of above- and below-ground plant parts (see Table 3.12). A reduction of almost 57% in the mass of the aerial part (stem and leaves) was observed after administration of salt stress. A net decrease of 60% in the underground mass of the plant, i.e. the roots, was also recorded after stress. The administration of the algal extract limited the decrease observed for the aerial parts to 46%, compared to 57% in the absence of algal extract. On the contrary, no effect on root mass loss was recorded after the application of the algal extract. Moreover, the administration of a volume of 2.25 mL of extract was the only procedure that seemed to have a slight positive effect towards the impact of salt stress. More generally, salt stress induced a 50% loss in mass of all plant tissues in avocado plants (Bonomelli et al. 2018).

Treatment	Aerial part (g of dry matter)	Subterranean part (g of dry matter)
Without salt stress	155	48
Salt stress	67	19
Salt stress + 1.5 mL ANE	78	15
Salt stress + 2.25 mL ANE	83	19

Table 3.12. *Effects of the application of* Ascophyllum nodosum *extract on the above-ground and below-ground parts of avocado plants subjected to salt stress (from Bonomelli et al. 2018)*

The application of *A. nodosum* extract does not show a notable effect on the growth criteria of plants subjected to salt stress over the culture period (240 days; Bonomelli et al. 2018). In contrast, it shows a protective effect on the height of the stressed plant during the first 30 days following the stress (see Figure 3.15). Indeed, over this period, the height of the unstressed plant (102 cm) is close to that of the stressed plant treated with 2.25 mL of algal extract (100 cm). For comparison, the height of the stressed but untreated plant is about 82 cm 30 days after salt stress (−19.6%). The height of the stressed but treated plant then varies little throughout the growing period and will be exceeded by that of the unstressed plant as early as the 60th day. In spite of this, the height of the stressed but treated plant always remains higher than that of the plant subjected to salt stress without treatment, and this during the whole culture cycle.

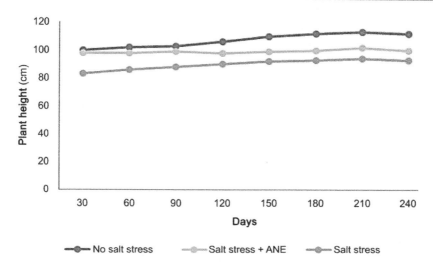

Figure 3.15. *Effect of salt stress on avocado plant height with and without* Ascophyllum nodosum *extract (ANE) supplementation (from Bonomelli et al. 2018). For a color version of this figure, see www.iste.co.uk/fleurence/algae.zip*

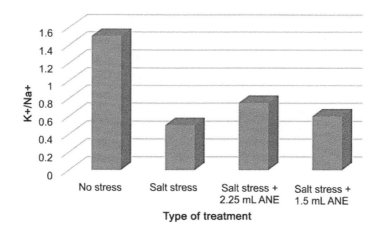

Figure 3.16. *Effect of salt stress with or without application of an extract* Ascophyllum nodosum *(ANE) on the K+/Na+ ratio in avocado leaves after 240 days of cultivation (from Bonomelli et al. 2018)*

In salt-stressed seedlings, a drastic decrease in the K^+/Na^+ ratio is observed (–66%; see Figure 3.16). This decrease was increased to 50% when the stressed plants were treated with algal extract. These decreases in the K^+/Na^+ ratio due to a decrease in intracellular potassium to the benefit of sodium, a "chaotropic" element, are indicators of the physiological disorder suffered by the plant during salt stress. These disorders concern the physiological processes controlled by potassium, such as turgor pressure or gas exchange occurring at the stomata.

The appearance of necrosis is also an indicator of the plant's response to stress, whether it is abiotic or biotic in origin. In the case of salt stress applied to avocado plants, the percentage of necrosis is increased by 300% compared to unstressed plants. The application of the algal extract treatment does not change this percentage, suggesting that the treatment has no effect on salt stress-induced leaf tissue necrosis (see Figure 3.17).

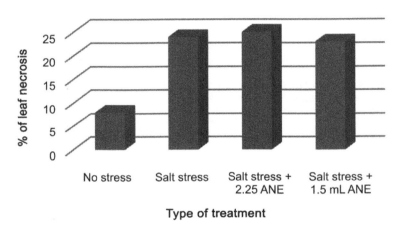

Figure 3.17. Effect of salt stress with or without the application of Ascophyllum nodosum extract

The effect of the application of an *Ascophyllum* extract on avocado plants subjected to salt stress thus appears to be limited to the preservation of the plant's height mainly in the first 30 days following the stress. However, the administration of the extract is insufficient to protect the plant from all the negative impacts generated by this type of stress, especially over the entire duration of the culture (240 days).

The effects of treatment with a commercial extract of *A. nodosum* (Acadian Seaplants) on the development of salt stress tolerance when growing eggplant (*Solanum melongena* L.) are also well documented. The stress is caused by increasing NaCl supply (320, 3,200, 4,800 ppm) during the irrigation of cultures (Hegazi et al. 2015). The increase in salt content has a negative effect on plant growth, resulting in a significant decrease in fresh shoot biomass and root system. Treatment with *A. nodosum* extract at the concentration of 0.5% (v/v) proves protective. In particular, it eliminates the negative effects on shoot growth when a moderate salt stress (320 ppm) is applied. It also limits these effects in the case of exposure to higher salt concentrations (see Figure 3.18). On the development of the root system, a positive algal extract impact is mainly observed in the case of moderate stress (320 ppm). It is much more mitigated for higher saline concentrations (see Figure 3.19).

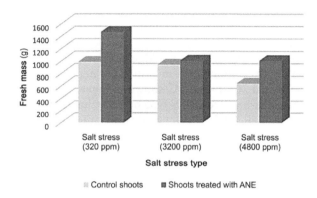

Figure 3.18. *Effect of salt stress with and without the application of* Ascophyllum nodosum *extract on the fresh mass of eggplant shoots (from Hegazi et al. 2015)*

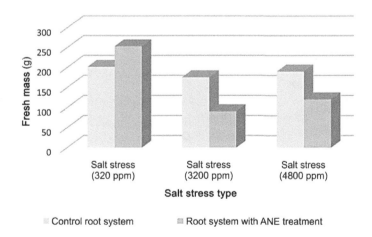

Figure 3.19. *Effect of salt stress with or without the application of* Ascophyllum nodosum *extract on the fresh mass of the eggplant's root system (from Hegazi et al. 2015). For a color version of this figure, see www.iste.co.uk/fleurence/algae.zip*

Other plant growth parameters such as plant height, number of branches, number and area of leaves are affected by salt stress. These negative impacts are significantly reduced by the application of the algal extract for moderate salt stress.

In the above case, it is therefore accepted that a commercial extract of *A. nodosum* strongly improves the tolerance of eggplant to moderate salt stress and partially limits the deleterious effects of higher stress.

Another study carried out on barley (*Hordeum vulgare* L.) subjected to a salt stress of 350 mM NaCl reports on the protective effect of an algal extract against this type of abiotic stress. The tested extract is obtained from the brown alga *Cystoseira mediterranea* collected on the Algerian coast (Bensidhoum and Nabti 2021). The extract obtained formulated in distilled water is tested in concentrated (100%) or diluted (50 and 20%) form. As shown in Figure 3.20, salt stress strongly inhibits germination of barley seeds. This type of inhibition is frequently cited in the literature (Levigneron et al. 1995). However, this inhibition of germination is lifted when the salt-stressed seeds were treated with a diluted extract of Cystoserum (20%).

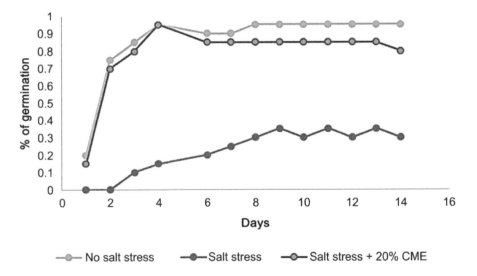

Figure 3.20. *Effect of salt stress with or without the application of diluted* Cystoseira mediterranea *extract (CME) on the germination of barley (*Hordeum vulgare L.*) grains (from Bensidhoum and Nabti 2021). For a color version of this figure, see www.iste.co.uk/fleurence/algae.zip*

As seen previously, plant growth is strongly affected by salt stress. The biomass corresponding to the aerial and root parts is generally reduced compared to that of unstressed plants. This is true for barley, but for this example, the cystoser extracts (20 and 50%) protect the plant from biomass loss (see Figure 3.21).

In addition to this protective effect, the algal extracts slightly improved the production of above-ground plant biomass, even compared to the control, the unstressed plant.

Salt stress has a very marked effect on the root system, decreasing the root biomass of the root system by nearly 67% compared to the root biomass of the unstressed plant (see Figure 3.22). The application of algal extracts significantly limits the impact of salt stress on root development.

Other more biochemical or physiological parameters affected by salt stress are better preserved after the application of diluted *Cystoserum* extracts. This is the case for chlorophylls a and b for which salt stress decreases production. This production is clearly improved after treatment of stressed plants with algal extracts (+37%).

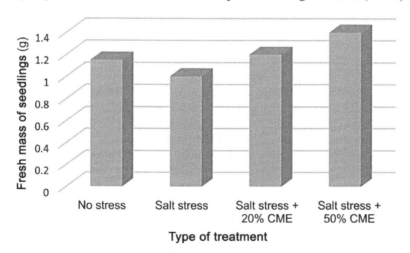

Figure 3.21. *Effect of salt stress with or without the application of diluted* Cystoseira mediterranea *(CME) extracts (20 and 50%) on fresh biomass of seedlings (aerial part) (*Hordeum vulgare L.*) (from Bensidhoum and Nabti 2021)*

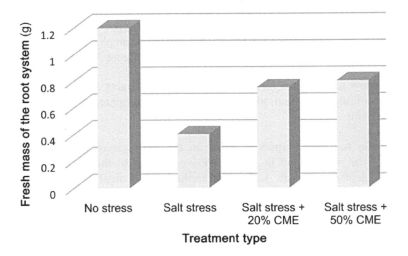

Figure 3.22. *Effect of salt stress with and without the application of diluted Cystoseira mediterranea extracts (CME) (20 and 50%) on the fresh biomass of the root system (*Hordeum vulgare L.) (from Bensidhoum and Nabti 2021)*

The Arjuna (*Terminalia arjuna*) is an ornamental tree native to India and Sri Lanka. Its bark and leaves are used as a decoction in traditional Indian medicine (Ayurveda). The effects of applying a commercial "Super Bluegreen" extract on the tolerance of Arjuna to significant salt stress (6,000 ppm NaCl) were the subject of a recent study (Abdel-Twab et al. 2020). Although referred to as "algal extract" by the authors, the commercial name of the tested product refers to the cyanobacterium *Aphanizomenon*, often commercialized under the name "blue-green algae", just like spirulina (Fleurence 2022). Foliar spraying of the commercial extract gave very good results in preserving photosynthetic pigment parameters (Chlorophylls a, b) in Arjuna plants subjected to salt stress (see Figure 3.23). The effects of the extract are mainly dose-dependent and are particularly effective with an extract at a concentration of 0.75% (w/v). They are less effective with lower concentrations (0.25 and 0.50%, Abdel-Twab et al. 2020). Optimal effects are also observed when the salt concentration of 6,000 ppm is reached with irrigation of the plants at four-day intervals.

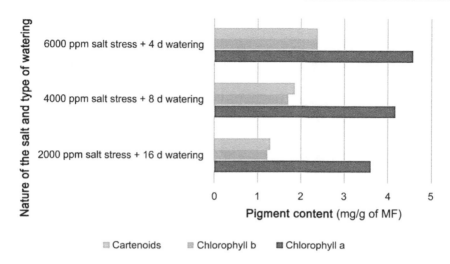

Figure 3.23. *Effect of the application of* Aphanizomenon *extract (0.75%) on chlorophyll and carotenoid pigment content of Arjuna subjected to increasing salt stresses (2,000, 4,000, 6,000 ppm NaCl) (from Abdel-Twab et al. 2020). For a color version of this figure, see www.iste.co.uk/fleurence/algae.zip*

Proline, an osmoregulatory amino acid, tends to accumulate in leaves under salt stress (Belkhodja 1996; El-Iklil et al. 2002). The application of the extract at the concentration of 0.75% (w/v) very significantly reduces the proline content at the leaf level (–20%) indicating a decrease of stress in the plant. Apart from these biochemical criteria, a similar observation was made with vegetal criteria such as plant height. The latter is maximal when the sprayed extract is at the concentration of 0.75%, a salt stress of 6,000 ppm and a six-day irrigation are applied (see Figure 3.24).

This study shows that a commercial cyanobacteria-based extract administered by foliar spray limits the impacts of salt and water stress on Arjuna seedlings, a tropical tree often subjected to both types of stress.

Some molecules widely distributed in macro- and microalgae have also been identified for their protective effect against salt stress (Carillo et al. 2020). These are mainly carbohydrates of tetrahalose type present in microalgae of the genera *Chlamydomonas* or *Chlorella* (Carillo et al. 2020) or the organosulfur derivatives of the zwitterion type (see Figure 3.25) present in the whole algal kingdom. In particular, 3-dimethylsulfuniopropionate is described as an effective osmoprotector against the impact of salt stress on the plant (Carillo et al. 2020).

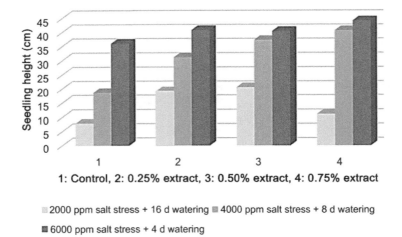

Figure 3.24. *Effect of the application of* Aphanizomenon *extracts (0.25%; 0.50; 0.75%) on the height of Arjuna seedlings under increasing salt stress (2,000, 4,000, 6,000 ppm NaCl) (from Abdel-Twab et al. 2020)*

In general, salt stress induces many morphological, anatomical and biochemical disorders in the stressed plant. Treatment with algal extracts, whether macro- or microalgae, significantly stimulates the biochemical defense mechanisms of the plant to this type of stress (see Table 3.13). These mechanisms involve the activation of enzymes that are classically mobilized to respond to abiotic stress. These are mainly catalase (CAT), superoxide dismutase (SOD), glutathione reductase (GR), ascorbate peroxydase (APX) or peroxidase (POD).

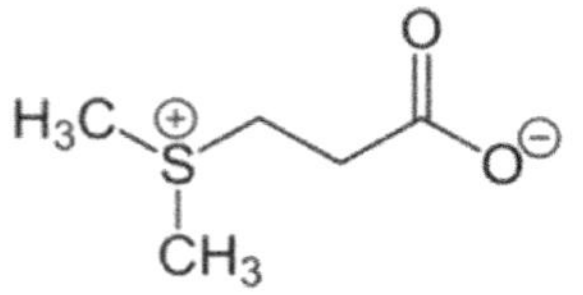

Figure 3.25. *Structure of 3-dimethylsulfuniopropionate*

The induction of the biochemical response by algal extracts is not limited to the increase of enzymatic activities involved in the protection phenomena with regard to oxidation processes generated by salt stress.

The production of photosynthetic pigments such as chlorophylls or secondary pigments such as carotenoids also seems to be activated via the treatment of the stressed plant with an algal extract (see Table 3.13). Other compounds such as phenols or tannins, molecules whose synthesis activation is part of the biochemical defense mechanisms of plants facing stress, see their production increase under the effect of algal treatment.

As shown in Table 3.13, different species of macroalgae, microalgae or cyanobacteria are likely, through the extracts produced from them, to generate salinity tolerance in many crops of agronomic interest such as wheat.

Algal species	Plant	Biochemical mechanisms at the basis of tolerance
Ascophyllum nodosum *	Eggplant (*Solanum melongena*)	Increase in SOD and APX activity, phenol and tannin content, K^+/Na^+ ratio
Ascophyllum nodosum *	Sprenger's asparagus (*Asparagus aethiopicus*)	Increased phenol and chlorophyll content and antioxidant enzyme activities
Sargassum muticum *	Chickpea (*Cicer arietinum*)	Increase in photosynthetic pigments and antioxidant enzyme activities (SOD, CAT, APX, POD)
Ulva lactuca *	Common wheat (*Triticum aestivum*)	Increase in the activities of antioxidant enzymes (SOD, CAT, PA, GR)
Ulva rigida *	Durum wheat (*Triticum durum*)	Increase in antioxidant enzymatic activities
Dunaliella salina **	Tomato (*Solanum lycopersicum*)	Increase in phenolic compounds and antioxidant enzymatic activities (CAT, POD, SOD)
Phaeodactylum spp. **	Bell pepper (*Capsicum annuum*)	Reduction of superoxide radical production and lipid peroxidation
Scenedesmus obliquus **	Common wheat (*Triticum aestivum*)	Increase in antioxidant enzyme activities (SOD, APX, CAT, POD) and in chlorophyll and carotenoid pigment contents
Spirulina platensis ***	Common wheat (*Triticum aestivum*)	Increase in antioxidant enzyme activities (SOD, APX, CAT, POD) and in chlorophyll and carotenoid pigment contents
Spirulina maxima ***	Common wheat (*Triticum aestivum*)	Increased antioxidant activities and tocopherol and carotenoid contents

Table 3.13. *Effect of macroalgae (*), microalgae (**) and cyanobacteria (***) on the activation of the biochemical response of plants developing tolerance to salt stress (from Carillo et al. 2020)*

The genetic mechanisms implemented via the application of an algal extract, and more particularly *A. nodosum* to explain the induction of salt tolerance, are partially elucidated in the model plant *A. thaliana* (Jitesh et al. 2012; Fleurence 2022).

A. nodosum extract is thought to negatively regulate some genes involved in salt stress sensitivity. Its potential target would be the gene encoding for the inhibitor of pectin methylesterase (At1g62760, Jithesh et al. 2012). This inhibitor would indeed play a role in the plant's sensitivity to salt and any other abiotic stress. Its down-regulation achieved by mutation allows the appearance of phenotypically salt stress-tolerant seedlings, this phenomenon being reflected in particular by an increase in daily root growth and biomass (see Figure 3.26 and Table 3.14).

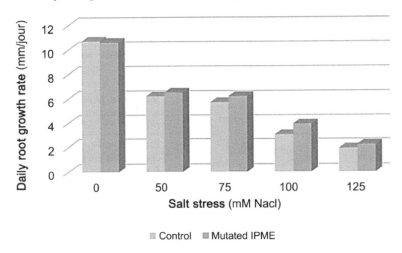

Figure 3.26. *Impact of salt stress on root growth of wild-type and mutated (IPME)* Arabidopsis thaliana *plants (from Jithesh et al. 2012). For a color version of this figure, see www.iste.co.uk/fleurence/algae.zip*

	Absence of salt stress	Salt stress (75 mM)	Salt stress (100 mM)
Biomass wild plant (g)	58	33	26
Biomass IPME mutated plant (g)	52	51	38

Table 3.14. *Effect of the IPME mutation on the growth of* Arabidopsis thaliana *in the presence of salt stress (from Jithesh et al. 2012)*

A more recent study also conducted on the model plant *A. thaliana* highlights another mechanism of action of *A. nodosum* extract on salt tolerance induction (Shukla et al. 2018). The algal extract would regulate the activity of microRNAs (miRNAs) involved in the expression of genes playing an important role in growth, development and stress tolerance. In particular, it would decrease the expression of ath-miR396-5p, ath-miR211b miRNAs and increase that of other RNAs such as ath-miR842 (see Figure 3.27).

This mechanism of genetic regulation based on the under- or overexpression of microRNAs involved in salt stress tolerance by *A. nodosum* extract represents one of the most advanced explanations on the subject today.

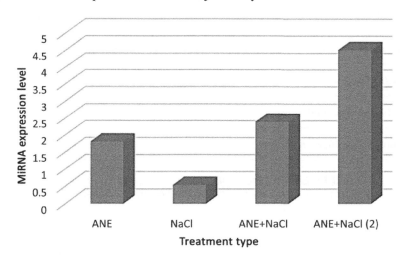

Figure 3.27. *Effect of* Ascophyllum nodosum *extract (ANE) on the expression of ath-miR842 microRNA from* Arabidopsis thaliana *subjected to a single salt stress (NaCl) and two consecutive salt stresses (NaCl (2)) (from Shukla et al. 2018)*

Algal extracts and more particularly brown algal extracts, such as *A. nodosum*, are inducers of the physiological, biochemical and genetic response of plants to salt stress. They strongly limit the impacts of moderate stress on plant growth. However, this effect is moderated by the duration of the stress and varies according to the salt concentrations experienced by the plant. Nevertheless, treatment with algal extracts confers to many crops of agronomic interest a relative tolerance to salt stress.

3.4. Tolerance to thermal stress

Abrupt or excessive temperature variations are abiotic stresses that can severely limit the yield of many agricultural productions. Cold weather followed or not by frost is a disabling factor for many crops. In the same way, heat, which can be accompanied by drought if it persists, represents a major abiotic stress for the culture of many plants.

The application of algal extracts, particularly *A. nodosum*, improves the tolerance of some crops to heat stress (see Table 3.15).

Seaweed extract or commercial product	Activity	Culture
Ascophyllum nodosum	Heat tolerance	Creeping bentgrass (herbaceous grass)
Ascophyllum nodosum	Improved frost resistance	*Arabidopsis*
Seasol (*Ascophyllum nodosum*)	Improved frost resistance	Vine
Maxicrop (*Ascophyllum nodosum*)	Improved frost resistance	Barley

Table 3.15. *Effect of the application of* Ascophyllum nodosum *extracts on the tolerance of selected crops to heat stress (from Chatzissavvidis et al. 2014 and Sharma et al. 2014)*

The mechanisms involved in the acquisition of this tolerance have been elucidated in some plants. Work on the model species *A. thaliana* has demonstrated the involvement of the lipophilic fraction of *A. nodosum* in the improvement of frost resistance (Rayirath et al. 2009). At the temperature of $-7.5°C$, the survival rate of plants not treated with the algal extract was 0%. The same survival rate (0%) is observed at the temperature of $-10°C$ for plants initially treated with lipophilic extract obtained from *Ascophyllum*.

The treatment with lipophilic algal extract therefore significantly improves cold tolerance ($+2.5°C$ negative). Treated plants show less leaf damage compared to those observed for plants treated with less. Chlorophyll losses were 70% lower during the freezing period than those reported for untreated plants. This finding appears to be the consequence of reduced expression of the AtCHL1 and AtCHL2 genes coding for Chlorophyllases.

Apart from this finding, *Ascophyllum* extract seems to play a modulatory role in the expression of the cold response genes COR15A, RD29A and CBF3. Under the effect of the lipophilic algal extract, the transcription of these cold tolerance genes is increased between 1.8 and 2.6 times (Rayirath et al. 2009).

Similarly, in tomatoes (*Lycopersicum esculentum*), a recent study provided a more precise characterization of the molecular mechanisms involved in heat stress resistance (Carmody et al. 2020). Two commercial extracts of *A. nodosum* (C129, PSI 194) are applied by foliar spray. The C129 extract is obtained by enzymatic treatment of the algal polysaccharide fraction and therefore potentially composed of oligosaccharides. The extract PSI-494 is made from the polysaccharide fraction via extraction at high temperature and under alkaline conditions.

Tomato plants during their growing cycle are subjected to a moderate stress (24 to 31°C) over a limited period of 14 days. The PSI-194 extract induces a tolerance to thermal stress clearly superior to that reported with the C129 extract. Its application significantly improved pollination, flowering and fruiting of treated plants compared to control plants.

Specifically, PSI-194 extract increases the transcription of HSP101.1, HSP70.9 and HSP17.7C-Cl genes (see Figure 3.28).

This mode of action appears to be the molecular mechanism involved in the acquisition of tolerance to moderate heat stress observed for this species.

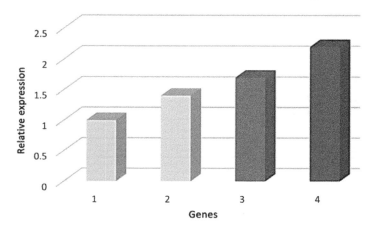

Figure 3.28. *Effect of* Ascophyllum nodosum *PSI-494 extract on the expression of HSP 101.1, HSP 70.9 and HSP 17.7C-Cl genes involved in heat stress tolerance (1: untreated plants; 2, 3, 4: treated plants. 2: HSP 17.7C-Cl gene; 3: HSP 70.9 gene; 4: HSP 101.1 gene) (from Carmody et al. 2020)*

Algal extracts and more particularly those of brown algae belonging to the genera *Ascophyllum* or *Laminaria* are often described as inducing mechanisms of tolerance to abiotic stress in plants (Fleurence 2022). These properties, associated with those stimulating plant growths, allow us to consider the use of algae as crop auxiliaries favoring the development of a more environmentally friendly agricultural production.

3.5. The quality of the products

The use of seaweed extracts also has a positive impact on the quality of agricultural products. This has been established for many crops such as vegetables or fruits.

Thus, on tomato (*S. lycopersicum*), the foliar application of a commercial product (Expando) with algae and yeast as its basis significantly improved the nutritional composition of the fruit for certain micronutrients and polyunsaturated fatty acids such as roughanic acid (+329%) (see Table 3.16; Mannino et al. 2020).

Mineral micronutrients (mg/100 g of fresh matter)	Untreated plants	Plants treated with Expando
Phosphorus	113.80	126.60
Magnesium	20.57	21.02
Iron	0.50	0.58
Manganese	0.15	0.16
Copper	0.09	0.12
Zinc	0.09	0.11
Unsaturated fatty acids	Untreated plants	Plants treated with Expando
C16:3 omega-3 (roughanic acid) (mg/100 g of fresh matter)	0.202	0.866
C16:1 omega-7	0.431	0.655
C16:1 omega-10	0.089	0.155

Table 3.16. *Effect of the application of a commercial seaweed extract (Expando) on the composition of micronutrients and some unsaturated fatty acids on tomato (*Solanum lycopersicum*) plants (from Mannino et al. 2020)*

Foliar application of a red algal extract, obtained from the species *Kappaphycus alvarezii*, on plants of the tomato variety (*Lycopersicon esculentum*) shows a positive effect on fruiting and fruit quality (Zodape et al. 2011). Spraying an extract with the concentration of 5% algae significantly improves the diameter (+25–32%) and ascorbic acid concentration of the fruits (see Figure 3.29). The flesh content of the fruits is also increased after treating the plants with algal extract (+31%). The macro- and micronutrient composition, representative of the nutritional value of the tomato, is clearly improved by the pre-treatment of the plants with the algal extract (see Table 3.17). Most of the observed beneficial effects are optimal after the application of algal extract at the 5% concentration. For lower or higher concentrations, a decrease in the intensity of these effects is noted, especially for fruiting yield and ascorbic acid composition (see Figure 3.29).

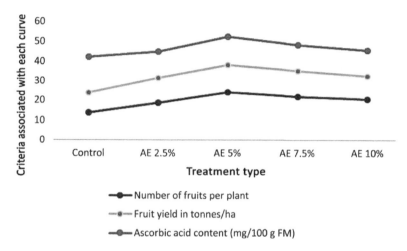

Figure 3.29. *Dose effect of an algal extract (AE) of* Kappaphycus alvarezii *on fruiting of tomato plants of the* Lycopersicon esculentum *variety (from Zodape et al. 2011). For a color version of this figure, see www.iste.co.uk/fleurence/algae.zip*

	N (%)	P (%)	K (%)	Fe (mg/kg)	Cu (mg/kg)	Zn (mg/kg)	Mn (mg/kg)
Witness	3.70	0.42	2.00	70.80	5.95	30.20	35.20
AE treatment (5%)	4.19	0.53	3.35	91.40	7.45	37.40	50.20

Table 3.17. *Effect of the application of an algal extract (AE) of* Kappaphycus alvarezii *on the mineral composition of tomatoes of the* Lycopersicon esculentum *variety (from Zodape et al. 2011) (NB: expressed on dry matter)*

Algal extract	Treated crop	Vegetable, fruit or derived product	Improved quality criteria
Ascophyllum nodosum	Spinach	Spinach	– Conservation – Flavonoid content – Nutritional value
Ascophyllum nodosum	Olive tree	Olive oil	Increase in linolenic and oleic acid content
Ascophyllum nodosum	Vine	Grapes	Increase in the anthocyanin content of the skin
Ascophyllum nodosum (*Cytozyme*)	Pomegranate tree	Pomegranate	– Reduction of the cracking phenomenon – Increase in fruit size
Ascophyllum nodosum	Cherry tree of staccato variety	Cherry	– Increase in fruit size – Increased meat content – Decrease of the acidity of the fruit – Increased polyphenol and vitamin C content
SWE (Seaweed extract)	Okra	Fruit of the okra	– Improvement of the nutritional value of the fruit by increasing the content of fiber, protein, sugars, vitamin C, phosphorus, calcium and magnesium
Chlorella vulgaris + bacteria extract	Romaine lettuce	Romaine lettuce	– Significant increase in size (+19%) – Increased carotenoid and antioxidant content

Table 3.18. *Effect of the pre-harvest application of macro- or microalgal extracts (*Chlorella vulgaris*) on the quality of some products from vegetable or fruit crops (from Abubakar et al. 2013, Battacharyya et al. 2015, Kopta et al. 2018, Gonçalves et al. 2020, Swarnam et al. 2020, Frioni et al. 2021)*

The effect of algal extracts on plant root development is well established in scientific literature (see section 3.1). The improvement of rhizogenesis by increasing the soil nutrient uptake capacity of the treated plants has a consequence on the mineral content of the latter. This mechanism is probably at the origin of the mineral enrichment observed in the fruits of plants subjected to algal treatment.

Apart from tomato, the impact of the application of algal extracts on the quality of other products from vegetable or fruit crops is also well known. The commercial extracts applied are mainly obtained from brown algae belonging to the *A. nodosum* species (see Table 3.18). Applications are made at the crop level and mostly by foliar spraying. The improvements in product quality concern criteria as varied as the biochemical composition of the products, their size, their diameter and their color via an increase in pigment content.

Numerous studies have reported the beneficial effect of treating crops with algal extracts on the quality of the fruits produced. The treatment of apple trees (*Malus domestica* of the Jonathan variety) with *A. nodosum*-based algal extract (4 kg/ha) very significantly improves the red color intensity of harvested apples (Soppelsa et al. 2018). This is mainly due to the increase in the anthocyanin content of fruits from trees subjected to algal extract treatment.

This treatment also showed a positive effect, although slight, on the weight (+5%) and size (+1.7%) of the fruit obtained. However, the most spectacular effect of the algal treatment is observed with respect to a disorder that affects this apple variety during its post-harvest storage. This disorder is linked to the appearance of small brown blisters that make the fruit more difficult to market. This disorder, which affects the visual aspect of the fruit, is called "Jonathan spot" and can be attributed to a fruit disease.

Figure 3.30. *Post-harvest "Jonathan spot" disease on the Jonathan apple variety (photo credit © A. Davis, Museums Victoria 1949, CC BY 4.0). For a color version of this figure, see www.iste.co.uk/fleurence/algae.zip*

Trials carried out on other apple varieties (Jonagold Decosta, Golden Delicious, Gala Must and Elstar) show the beneficial effect of seaweed-based products on the quality of the fruit produced. However, the effects induced can be moderated by seasonal factors or by the nature of the extract used.

Experiments carried out by foliar spraying of commercial products (Kelpak and Goëmar BM86) have demonstrated this phenomenon on the above-mentioned apple varieties (Basak 2008). The treatments are carried out from flowering and stopped four weeks before harvest.

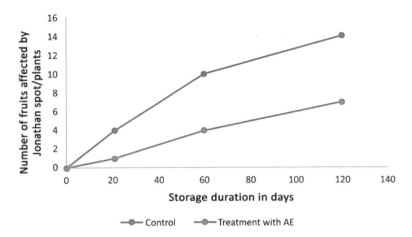

Figure 3.31. *Effect of crop treatment with algal extract (AE) on the number of fruits affected by "Jonathan spot" during the fruit storage period (from Soppelsa et al. 2018). For a color version of this figure, see www.iste.co.uk/fleurence/algae.zip*

The effects induced by these treatments on fruit number, weight, diameter and the red color of the fruit differed depending on the apple variety, the year of production and the algal extract used.

This is particularly observed on the Gala variety Must (see Figures 3.32 and 3.33). On this variety, foliar application of Kelpak resulted in a gain in fruit growth of nearly 35% during the 2003 production. The following year, the application of the same product did not result in a significant improvement in the weight of harvested fruit. In 2003, the application of Goëmar BM86 extract induced a 20% gain in fruit weight compared to the control production. In 2004, no significant difference in fruit weight was observed between the control production and the one treated with Goëmar extract. These results show a beneficial effect of the application of the seaweed extracts on this variety, weighted, however, by a seasonal agronomic impact.

Similarly, the application of the commercial extracts mentioned above has a clear influence on the red coloration of the fruits (see Figure 3.33). This impact is particularly noticeable for fruits from the 2003 harvest. As previously discussed, this result highlights the effect of the seaweed extracts on the quality of the harvested fruits. For the Gala must variety, it confirms the importance of the seasonal factor on the effectiveness of Kelpak or Goëmar.

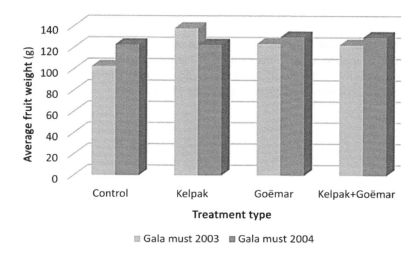

Figure 3.32. *Seasonal effect of the treatment of Gala Must apple trees with commercial seaweed extracts (Kelpak, Goëmar BM 86) on the weight of harvested fruit (from Basak 2008). For a color version of this figure, see www.iste.co.uk/fleurence/algae.zip*

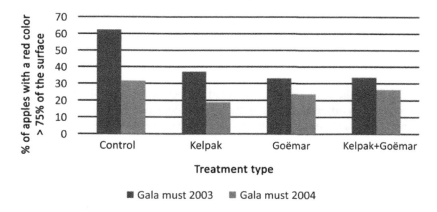

Figure 3.33. *Seasonal effect of the treatment of Gala Must with commercial seaweed extracts (Kelpak, Goëmar BM 86) on the red color of harvested fruits (from Basak 2008). For a color version of this figure, see www.iste.co.uk/fleurence/algae.zip*

The impact of the application of seaweed extracts on the quality of fruits after harvest and during storage is also known. A study conducted on oranges of the

Valencia variety shows a beneficial effect of the application of a commercial product (Cytolan Star) on fruit preservation during cold storage (Kamel 2014). The commercial algal extract mainly composed of brown algae, including the species *A. nodosum*, is applied at the concentration of 1 g/L. When combined with garlic oil (0.5%), the activity of the resulting mixture very strongly limits the percentage of fruit attacked by rot after 60 days of storage (see Figure 3.34).

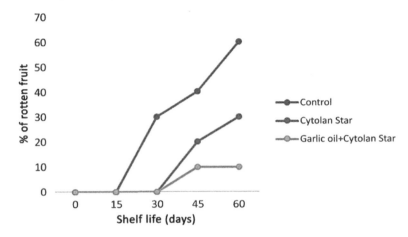

Figure 3.34. *Effect of the application of a commercial algae-based product (Cytolan Star) on delaying the onset of rot on oranges of the Valencia variety stored in cold storage for 60 days (from Kamel 2014). For a color version of this figure, see www.iste.co.uk/fleurence/algae.zip*

The effects of the application of seaweed extracts on the post-harvest quality of fruits or vegetables differ according to the species considered. On eggplant, avocado or pear, no significant impact on ripening speed was observed after immersion in an algal extract (Algistim, Blunden et al. 1978). On the contrary, the shelf life of peppers was significantly improved in the presence of commercial extracts (SM3 or Marinure, Blunden et al. 1978). The most remarkable effects are noted for limes. For limes, immersion for one hour in the commercial extract Marinure or SM3 at a concentration of 30 ppm delayed the onset of "graying" phenomena for up to 34 and 47 days respectively. This duration of preservation of the fruit color is clearly greater than that observed for the control fruits which start to yellow as early as 12, or even 16 days, depending on the storage conditions.

In general, the use of seaweed extracts with regard to fruit quality does not show negative effects on fruit quality. Depending on the species treated and the year of production, it can generate negligible or positive effects. This course of action is more in line with an integrated approach to the use of algae in agriculture. Such an approach, which would go from plant growth to the qualitative production of food generated by the crops, would fully meet the expectations of agrobiology.

The use of algae generally has a positive impact on plant growth. The mechanisms of action are numerous and varied (improvement of the soil structure, activation of the rhizobacterial flora, contribution of nutrients and or phytohormones). According to the doses in use, they are considered as fertilizers or biostimulants. The use of algae in the treatment of crops also improves the tolerance, or even the resistance, of plants to abiotic stresses as well as to aggression by pathogens (see section 5.1).

Finally, and under certain conditions, the addition of seaweed extracts before or after harvest can show a beneficial effect on the quality of the fruits and their conservation duration. For most of the reported effects, brown algae are mainly mentioned (*Ascophyllum* sp., *Ecklonia* sp.). This translates into the presence of these species in the majority of marketed products.

In conclusion, seaweeds are very interesting biological inputs to promote the development of an organic agriculture or agrobiology.

4

Feeding of Livestock

Algae can also be used as feed for livestock. A livestock animal, or production animal, is an animal that is raised for its economic profitability. It is used for the production of food, hides or for other agricultural purposes (e.g. clearing, plowing). Historically, this term was applied to domesticated land animals consumed by humans and/or involved in agricultural work. Aquatic animals produced for food purposes (fish, mollusks, crustaceans) and generating economic profits also meet this definition of livestock.

Multicellular algae or macroalgae have often been included in the diet of some ruminants (cattle, sheep). Unicellular algae or microalgae, as well as cyanobacteria, are most often included in the diet of filter-feeding mollusks and certain crustaceans. They are also associated with the diet of fish via feeding on their rotifer prey. Independently of this, some species of gastropod mollusks, such as abalones, are reared on a diet with macroalgae as its basis belonging to the genera *Enteromorpha* or *Palmaria*.

4.1. Ruminant nutrition

Numerous examples of algae consumption by ruminants (sheep, goats, cattle) are reported in historical (see Chapter 1) or scientific literature.

In the north of Scotland in the Orkney Islands, sheep frequently feed on seaweed washed up on the foreshore (see Figures 4.1 and 4.2).

Figure 4.1. *Geographical location of the Orkney Islands (from Google Earth). For a color version of this figure, see www.iste.co.uk/fleurence/algae.zip*

This coastal grazing practice is found to be widespread in northern European islands (UK, Iceland, Scandinavia), given the limited land available for extensive livestock production (Delaney et al. 2016).

Figure 4.2. *Orkney Islands sheep grazing on the foreshore (photo credit © Orkney.com, 2021). For a color version of this figure, see www.iste.co.uk/fleurence/algae.zip*

Sheep under this type of grazing consume a wide variety of algae. They show a preference for brown algae of the species *Laminaria digitata* and *L. hyperborea* in summer (Hansen et al. 2003). An experiment conducted on sheep previously fed grass for five months and subsequently fed a diet of a mixture of Laminaria (3–5 kg of *L. hyperborea-L. digitata*) shows a daily algal intake by the animals of about 1.4 +/–0.2 kg (Hansen et al. 2003). This level of intake was not significantly different from that of the control group of sheep fed exclusively with the algal mixture.

The digestibility rates of dry matter and organic matter were respectively 71.7% and 79.6% over 48 hours with the group previously fed grass and subsequently subjected to the algal diet.

The group initially not adapted to the consumption of seaweed was able to digest this type of food. Laminaria, although rich in anti-nutritional factors (polyphenols), are therefore easily digestible by these ruminants. As such, they appear as suitable substitutes for the nutrition of some sheep.

	Herbaceous food (% total ciliates)	Seaweed-based diet (% total ciliates)
Dasythrica ruminantium	3.1	36.5
Isotricha prostoma	4.8	1.9
Isotricha intestinalis	0.2	–
Entodinium spp.	89.3	61.6
Polyplastron multivesiculatum	2.4	0.1

Table 4.1. Biological diversity of the ciliated protozoan flora contained in the rumen of sheep which are fed a herbaceous or algae-based diet (from Orpin et al. 1985)

A study of the gut flora of Orkney sheep reveals that it is close to that of other ruminants feeding on terrestrial plants. The population of ciliated protozoa is quantitatively comparable in the rumen of both types of ruminants (Orpin et al. 1985). In contrast, in animals feeding exclusively on algae, the species *Dasytricha ruminantium* appears to be the dominant ciliate species in the rumen. It represents nearly 37% of the ciliate population (see Table 4.1). The biodiversity of the bacterial flora is quite similar in the two rumen types. Despite this, sheep feeding on algae show a content of *Oscillospira guilliermondii* 10 times higher than that detected in the rumen of animals fed with grass. Other bacteria such as *Streptococcus bovis* or *Selenomonas ruminantium* are also better represented in the rumen of animals which are fed an algal diet (Orpin et al. 1985).

Sheep in Orkney do not seem to differ significantly in their intestinal flora from that reported for ruminants feeding on a herbaceous pasture. On the other hand, the representativeness of the different species of the ciliated and bacterial flora differs according to the diet to which the sheep are subjected.

On the other hand, the bacterial flora isolated from the rumen of sheep fed with algae shows some ability to use algal polysaccharides for their development. This is less true for the bacterial flora from the rumen of sheep on grass pasture (see Figure 4.3).

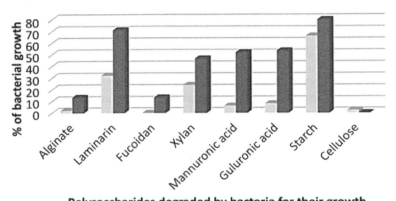

Polysaccharides degraded by bacteria for their growth

■ Bacterial flora algae-free diet   ■ Bacterial flora diet with algae

Figure 4.3. *Effect of selected algal polysaccharides on the bacterial growth of rumen flora isolated from animals which are fed an algae-free diet or algal diet (from Orpin et al. 1985)*

The efficiency of the bacterial flora of Orkney sheep to produce biomethane by fermentation of algal polysaccharides has also been the subject of a particular study. Isolates of rumenic bacterial microbiota are tested in vitro for their ability to degrade a crude extract of *Laminaria hyperborea* and a wide variety of polysaccharides including alginates (Williams et al. 2012). Of 65 isolates tested, nine degrade more than 90% of a polysaccharide such as laminarin (see Table 4.2). Depending on the isolate, hydrolysis of alginates or fucoidans was found to be partial to very partial. Isolate L7 is the most efficient in the degradation of laminarin, alginates, as well as *Laminaria hyperborea* extract. It is mainly represented by bacteria belonging to the genus *Prevotella*. The rumenic microbiota is also efficient in producing acetate and methane from frequently consumed algae (*Fucus* sp., *Ascophyllum nodosum, L. hyperborea*) on the foreshore by Orkney sheep. This observation allows a cross-valorization between two biological resources that are the microbiota of sheep and the production of bioethane by algae.

Rumen bacterial isolate	Laminarin (%)	Alginates (%)	Fucoidans (%)	Extracted from *L. hyperborea* (%)
L7	95.2	80.0	18.6	52.4
L8	93.5	66.7	8.5	19.0
L10	94.1	–	20.3	44.0
L12	58.2	5.8	1.7	14.5
A9	93.2	70.0	1.7	32.1
A11	93.6	71.1	8.5	24.9
A12	92.7	70.6	–	28.6
A14	93.6	71.1	–	22.6
C8	93.7	71.1	–	29.8

Table 4.2. *Degradation efficiency (%) of algal polysaccharides and Laminaria extract by rumen bacterial isolates from Orkney sheep (from Williams et al. 2012) (L: Laminarin, A: Alginate, C: Cellulose)*

The impact of an algae-based diet on the zootechnical performance of sheep and on the production of fermentative gases is also known for other species not belonging to the brown algae family. The effects of a diet supplemented with *Ulva lactuca* on the growth of Niamey sheep are notably well documented (El-Waziry et al. 2015). The addition of 3–5% algae to the basal diet does not induce growth gains in animals fed this diet. Similarly, the addition of *Ulva* to the diet of ruminants shows no effect on the production of fermentative gases such as biomethane. This result is in contrast to that previously described with brown algae. However, it can be explained by the different nature of the sheep studied, and thus of their microbiota, as well as by the algal species considered. The latter shows a polysaccharide profile, therefore fermentable, and very different from that of brown algae. It is characterized in particular by an absence of alginates, laminarin and a more preponderant presence of cellulose, ulvans and pectins.

The addition of *Ulva lactuca* (20%) to the diet of ruminants belonging to the Merino species has allowed the nutritional value of these algal proteins to be better characterized. The latter appear less degraded than proteins from other sources (Arieli et al. 1993). This translates into a lower rumen nitrogen level than that reported via a diet of concentrated proteins or vetch hay (see Figure 4.4). This finding is supported by the rumen's ammonia level of 391 mg/L on the protein concentrate diet versus 311 mg/L on the ulva diet. Nitrogen excreted in urine is also impacted by diet. It is 10% lower with the algal protein diet. These characteristics allow *Ulva lactuca* to be classified as a low-energy but high-nitrogen feed. This finding is conducive to the use of this alga with other high energy but low nitrogen foods such as cereals (Arieli et al. 1993).

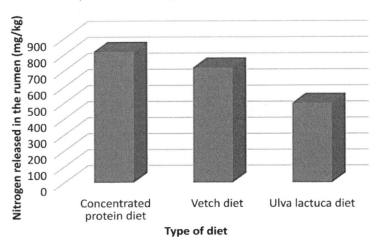

Figure 4.4. *Effect of some diets on nitrogen digestibility in the rumen of ruminants belonging to the Merino species (from Arieli et al. 1993)*

The value of using *Ulva lactuca* in goat diets has also been studied (Ventura and Castanon 1998). It was carried out on goats in the Canary Islands. The digestibility of algal organic matter in the rumen of goats was 59%, and the digestibility of crude protein was evaluated at 96 g/kg of algal dry matter. The observed degradation of the protein fraction in the rumen represented 54% of the protein supplied during the ingestion of the algal raw material.

In view of these results, ulva can be considered as a medium quality forage, but with a high protein content. This study on goats confirms the status of ulva as a complementary feed already reported for sheep by Arieli et al. (1993).

The same experiment is reported for cattle, which are ruminants of great economic interest in animal production. Different species of algae belonging to the brown, red or even green algae family are tested to evaluate the digestibility of their protein fractions in the rumen of dairy cows (Tayyab et al. 2016).

As shown in Figure 4.5, the highest digestibility of the protein fraction is observed with brown algae belonging to the genus *Alaria* (642 g/kg crude protein).

The algae of the genus *Ulva* show a significantly lower degradation of the protein fraction (237 g/kg crude protein).

These results are described for algae harvested in autumn. However, a seasonal effect is to be noted according to the species concerned and more particularly for those of the genus *Alaria*. For the latter, the digestibility of the algal protein fraction is 330 g/kg of crude protein in spring versus 642 g/kg in the fall.

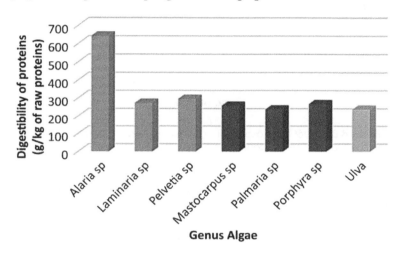

Figure 4.5. *In situ digestibility of the crude protein fraction of different algal species collected in the fall by dairy cows* (Alaria, Laminaria, Pelvetia: brown algae; Mastocarpus, Palmaria, Porphyra: red algae; Ulva: green algae) *(from Tayyab et al. 2016). For a color version of this figure, see www.iste.co.uk/fleurence/algae.zip*

The effect of algal intake on milk composition is also well informed through a study conducted in 2016 (Lopez et al. 2016). In the latter, a dietary supplementation based on the supply of *Ascophyllum nodosum* to dairy cows was followed to assess its impact on milk iodine concentration and microbiological quality. Animals fed a diet including the addition of seaweed produce milk with a significantly increased iodine concentration (see Figure 4.6). This result is particularly interesting, as it opens a perspective in the fight against iodine deficiency that affects many populations, including those living in European regions. Currently, this public health problem is mainly addressed by adding iodine to cooking salt. This study therefore opens up an alternative approach that is easier to implement towards categories of populations (e.g. pediatric populations) to limit the deleterious effects of dietary iodine deficiency.

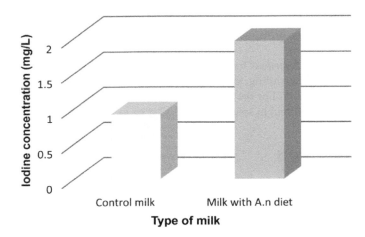

Figure 4.6. *Effect of* Ascophyllum nodosum *(A. n) in the diet of dairy cows on iodine concentration in milk (from Lopez et al. 2016)*

The diet with *A. nodosum* also had an impact on the milk microbiota. This results in a change in the representation of the different germs naturally present in the milk (see Table 4.3). The reduction in the presence of certain pathogenic germs such as enterococci or *Pseudomonas* appears to be a positive effect of the algal diet on the quality of the milk obtained.

Germs	Control milk (log cfu/g)	Milk diet *Ascophyllum nodosum* (log cfu/g)
Lactobacillus **mesophylls**	4.5	3.5
Lactococcus **mesophylls**	4.3	3.8
Enterococcal flora	3.6	2.4
Pseudomonas **spp.**	4.5	3.3
Total coliforms	3.9	3.3
Aerobic mesophyll flora	4.7	4.7

Table 4.3. *Effect of a diet supplemented with* Ascophyllum nodosum *on the lactational microbiota (from Lopez et al. 2016)*

Feeding of Livestock 77

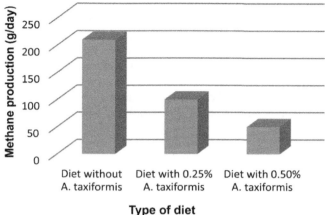

Figure 4.7. *Effect of integrating the red alga* Asparagopsis taxiformis *in the diet of cattle on enteric methane production (from Roque et al. 2021)*

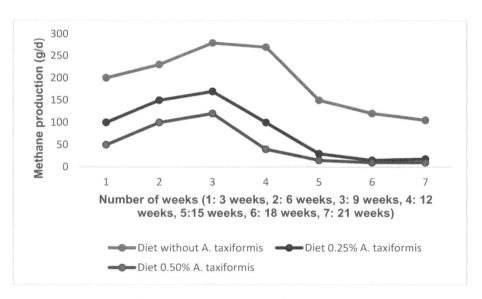

Figure 4.8. *Impact of feeding* Asparagopsis taxiformis *on daily enteric methane production over a rearing period of 21 weeks (Roque et al. 2021). For a color version of this figure, see www.iste.co.uk/fleurence/algae.zip*

However, the value of using an algal supplement in cattle feed is not limited to zootechnical criteria or the quality of the products produced. A recent study shows that the addition of the red alga *Asparagopsis taxiformis* in the nutrition of Angus cattle significantly reduces enteric methane production by these animals (Fleurence 2021b; Roque et al. 2021). This finding contrasts with results previously described for sheep fed with *Ulva lactuca*. As shown in Figure 4.7, methane production, a greenhouse gas, is reduced by a factor of four for animals fed this red alga (0.50% of the diet). This reduction is progressive and significant over the 21-week experiment (see Figure 4.8).

The incorporation of seaweed in the diet of the cattle does not affect the quality of the carcass and the organoleptic qualities of the Angus meat.

Similar results are presented in dairy cows. The addition of the red alga *Asparagopsis armata* in their diet at 0.5–1% of the feed ration significantly reduces enteric methane production by cattle (Roque et al. 2019). In contrast, a limited impact on CO_2 production is observed with the diet incorporating 0.5% algae. A significant difference, however, begins to emerge with the diet incorporating 1% algae in the diet (see Figures 4.9–4.11).

The production of hydrogen, a simple gas that is less controversial than methane for its greenhouse effect, is significantly increased as a result of the algae-based diet.

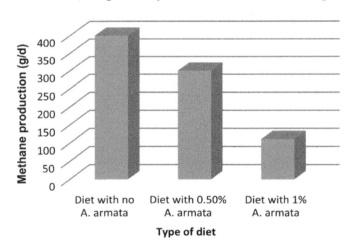

Figure 4.9. *Effect of the red alga diet* Asparagopsis armata *on enteric methane production by dairy cows (from Roque et al. 2019)*

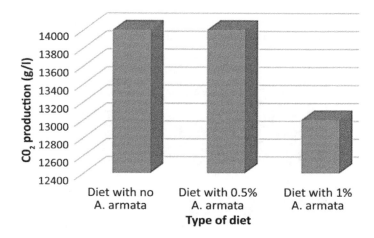

Figure 4.10. *Effect of the red alga diet* Asparagopsis armata *on the enteric production of carbon dioxide by dairy cows (from Roque et al. 2019)*

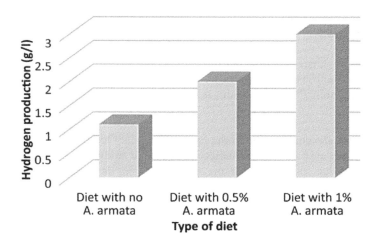

Figure 4.11. *Effect of the red alga diet* Asparagopsis armata *on enteric hydrogen production by dairy cows (from Roque et al. 2019)*

Nutrition with red algae of the genus *Asparagopsis* as its basis thus seems to be an interesting way to limit methane production from cattle. This approach appears to be an alternative way to curb the production of a greenhouse gas generated by cattle breeding. However, it is highly dependent on the bioavailability of the *Asparagopsis* species. Indeed, initially present on the coasts of Australia, New Zealand and

Reunion Island (see Figure 4.12), these species have only been established in the Northeast Atlantic and Mediterranean since 1925 (Cabioc'h et al. 1992).

Figure 4.12. *Red alga* Asparagopsis taxiformis *(Passe Jarre jarron, Marseille) (photo credit © O. Bianchimani, 2021). For a color version of this figure, see www.iste.co.uk/fleurence/algae.zip*

4.2. Pig nutrition

The use of algae in pig nutrition is a practice traditionally applied in Scandinavian countries (see Chapter 1). Recent studies have uncovered the impact of their use on animal growth, health and welfare (Ruiz et al. 2018; Corino et al. 2019).

In particular, the contribution of the feed supplement OceanFeed Swine (OFS) in the growth of piglets has been the subject of particular study (Ruiz et al. 2018). This commercial product developed for the feeding of pigs is composed of a mix of brown, red and green algae whose species are not specified by the company that produces it (Ocean Harvest Technology, Galway, Ireland[1]).

In piglets, the administration of this feed supplement is performed from weaning until the end of the fattening phase. The interest of this experimentation resides in the size of the sample involved (1,809 piglets) and in the rearing site, which is an industrial farm (Ruiz et al. 2018).

The seaweed-based feed supplement is provided at a rate of 5 g/kg of feed ration. This diet was fed to animals aged 21–160 days. The daily weight gain in the fattening phase was significantly improved with the supplemented diet (+3.2%).

---

1 https://oceanharvesttechnology.com/.

The OFS diet also appears to have a positive impact on the carcass weight of animals on this type of diet (see Figure 4.13).

The impact of brown seaweed extracts on the digestive flora of piglets and the iodine content of animal tissues is also well documented (Dierick et al. 2009). The experiment is based on the supply of dehydrated *Ascophyllum nodosum* to weaned piglets. It is performed both in vitro and in vivo. In vitro, the supply of the algae has an influence on the intestinal flora of the animals, decreasing the content of *Escherichia coli* in particular. This bacterium, common in the mammalian gut, can sometimes cause food poisoning. In vivo, the alga administered at a level of 10 g/kg of ration shows a reducing effect of the *E. coli* content in the stomach and small intestine. This decrease is correlated with an increase of the lactobacillus flora in the small intestine. The increase in the lactobacillus to the *E. coli* population ratio is considered to be a beneficial effect on the gut microbiota and the health of the animal.

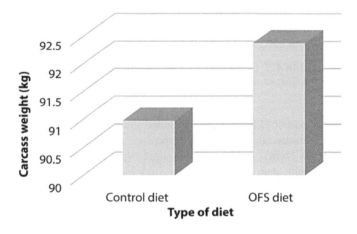

Figure 4.13. *Effect of the diet supplemented with OFS product on the carcass weight of animals slaughtered at the end of the of fattening (from Ruiz et al. 2018)*

Feeding seaweed also increases the iodine content of the muscles and many organs of the animal (see Table 4.4).

A similar experiment incorporating a *Laminaria* spp. extract into the diet of recently weaned pigs is also described (O'Doherty et al. 2010). In this experiment, the administration of the algal extract was coupled with the dietary supply of lactose at different concentrations. The resulting diets were supplied ad libitum for 25 days (see Table 4.5).

Tissues or organs	Algae-free diet	Diet with algae
Iliopsoas muscle	15.5	59.5
Longissimus dorsi muscle	19.9	55.4
Back fat	19.3	83.9
Liver	27.3	117.7
Kidney	31.4	214.3
Heart	29.3	134.3

Table 4.4. *Effect of Ascophyllum nodosum supplementation on the iodine content of different tissues and organs (μg/kg fresh mass) (from Dierick et al. 2009)*

Diet	Composition
T1	150 g of lactose/kg
T2	150 g of lactose/kg + algal extract
T3	250 g of lactose/kg
T4	250 g lactose/kg + algal extract

Table 4.5. *Composition of lactose-based diets fed ad libitum to weaned pigs (from O'Doherty et al. 2010)*

The Laminaria extract (2.8 g/kg diet) included in this experiment is characterized by laminarin and fucoidan content that are, respectively, 300 mg/kg and 276 mg/kg for diets T2 and T4. As shown in Figure 4.14, algal supplementation has a significant impact on the weight gain of pigs fed a high-lactose diet (T4). This gain is insignificant for animals fed a lower lactose diet (T1, T2). Regarding the digestive parameters, the intake of the algal supplement has an effect on the intestinal digestibility coefficient of organic matter, as well as nitrogen and energy release (see Figure 4.15). It very significantly increases these digestibility coefficients for the high-lactose diet (T4), but this increase is also observable for the low-lactose diet (T2).

The addition of seaweed to the high-lactose diet also has an influence on the intestinal flora of the animals. As previously described with *Ascophyllum nodosum* extract, *Laminaria* spp. extract limits the presence of *E. coli* in the gut flora in favor of *Lactobacilli* spp. (O'Doherty et al. 2010).

The inclusion of a seaweed extract, rich in laminarin and fucoidan polysaccharides, in a high-lactose diet therefore improves the growth performance of weaned pigs and the digestibility of the feed provided. It also reduces the representativeness of *E. coli* in the intestinal flora and in the feces. This diet is therefore particularly interesting to prevent the development of pathogenic flora in pigs.

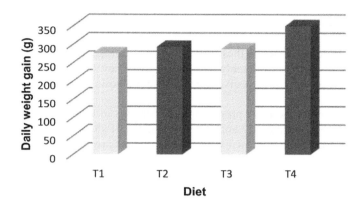

Figure 4.14. *Effect of a diet supplemented with* Laminaria *spp. extract on the daily weight gain of weaned pigs fed ad libitum for 25 days on a diet with varying lactose concentrations (T1: low-lactose concentration diet; T2: T1 +* Laminaria *extract; T3: high-lactose concentration diet; T4: T3 +* Laminaria *extract) (from O'Doherty et al. 2010)*

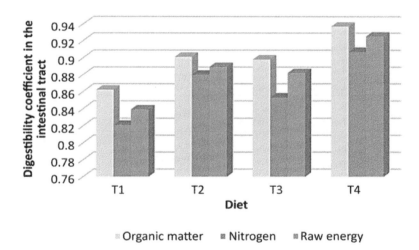

Figure 4.15. *Effect of a diet supplemented with* Laminaria *spp. extract on the digestibility coefficient of the intestinal tract of weaned pigs fed a diet with varying lactose concentrations (T1: low-lactose concentration diet; T2: T1 +* Laminaria *extract; T3: high-lactose concentration diet; T4: T3 +* Laminaria *extract) (from O'Doherty et al. 2010). For a color version of this figure, see www.iste.co.uk/ fleurence/algae.zip*

The presence of bioactive substances of a polysaccharide nature in brown, green or red algae seems to be the reason for the positive effects of giving algal feed to pigs (Corino et al. 2019). In brown algae, these are primarily laminarin and fucoidans in green algae, while these are ulvans and carrageenans for red algae. The positive effects observed, notably in piglets and sows, concern growth (see Table 4.6), improvement of digestive function and prebiotic activities.

Brown seaweed is the most promising feed for pigs in terms of immune system stimulation, antioxidant activity and digestive health. It should be noted, however, that studies on the effect of brown seaweed in pig diets are far more numerous than those reporting the impact of green or red seaweed-based nutrition.

Algal supplement	Dose	Animal	Daily weight gain
Ascophyllum nodosum	10–20 g/kg Dehydrated seaweed	Weaning pigs	+48.14%
Brown algae	50–100–200 mg/kg Oligosaccharides of alginic acid	Weaning pigs	+14.81% (50) +40.78% (100) +39.35% (200)
Ecklonia cava	0.10 g/kg Fucoidans	Weaning pigs	+6.98%
Laminaria digitata	0.314–0.250 g/kg Laminarin + fucoidans	Weaning pigs	+21.95%
Laminaria spp.	1 g/day Laminarin	Weaning pigs	+32.35%
Laminaria spp.	0.18–0.34 g/kg Laminarin + fucoidans	Weaning pigs	+16.30%
Laminaria spp.	(1–2–4 g/kg) Extract	Weaning pigs	+10.4% (1) +25.70% (2) +21.69% (4)
OceanFeed Swine	5 g/kg	Pigs (56–160 days)	+326%

Table 4.6. *Impact of adding laminarin-rich brown algae, fucoidans, alginates and a commercial seaweed product on the daily weight gain of pigs (from Corino et al. 2019)*

The antibacterial activities of laminarin are also the reason for the interest in supplementing the pig diet with brown seaweed extracts (see Table 4.7).

Bacterial strain	Algal species	Type of algae	Inhibiting molecules
Escherichia coli	– *Ascophyllum nodosum* – *Laminaria hyperborea*	Brown algae	Laminarin
Listeria monocytogenes	– *Ascophyllum nodosum* – *Laminaria hyperborea*	Brown algae	Laminarin
Salmonella typhimurium	– *Ascophyllum nodosum* – *Laminaria hyperborea*	Brown algae	Laminarin
Staphylococcus aureus	– *Eisenia bicyclis* – *Ascophyllum nodosum* – *Laminaria hyperborea*	Brown algae	– Phlorofucofuroeckol – Laminarin

Table 4.7. *Antibacterial activities of laminarin contained in brown seaweed extracts fed as a supplement to pigs in pig nutrition (from Corino et al. 2019)*

The effects of the dietary intake of laminarin and fucoidans on pork quality and preservation are also well known.

A study involving the administration of an extract enriched with laminarin (500 mg/kg ration) and fucoidans (420 mg/kg ration) provided in dehydrated (L/F-D) or wet (L/F-H) form to animals shortly before slaughter (21 days before) gives promising results on this subject (Moroney et al. 2012). These results relate primarily to the quality of meat stored at 4°C under a modified atmosphere (80% O_2, 20% CO_2).

As shown in Figure 4.16, meat from pigs fed with laminarin and fucoidan wet extract (L/F-H) shows better resistance to the meat rancidity process and its taste quality is therefore found to be better preserved during storage time.

On the other hand, other parameters such as the tenderness or the color of the meat during storage seem to be little affected by the administration of the algal polysaccharide supplemented diet. Similarly, this diet does not appear to have any effect on the microbiological quality of the meat stored in a modified atmosphere. This result does not call into question those previously obtained with these polysaccharides and which concerned the growth of pigs of different ages and the quality of their digestive microbiota.

However, the use of brown algae in pig feed can have an effect on the quality of the meat and especially on its color. Thus, the addition of the alga *Macrocystis pyrifera* as a feed supplement in the diet of pigs in their fattening phase shows an impact on the quality of the meat produced, particularly in terms of its color (redness of the flesh) (Jerez-Timaure et al. 2021). The latter seems to decrease with the percentage of algae added (see Figure 4.17).

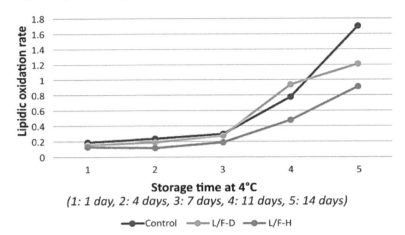

Figure 4.16. *Effect of the dietary supply of algal extracts, obtained from* Laminaria digitata, *concentrated in laminarin and fucoidans on pork rancidity (expressed as lipid oxidation rate or mg malondialdehyde/kg pork) (from Moroney et al. 2012). For a color version of this figure, see www.iste.co.uk/fleurence/algae.zip*

Studies carried out on pigs show that the integration of algae, and more particularly brown algae, generally has a positive effect on feed intake, animal growth and the quality of their microbiota. In addition, a positive impact on the digestive function of the animals is often reported. The polysaccharides contained in some brown algae probably play a role as prebiotics facilitating the development of an intestinal flora beneficial to the digestive health of the animal.

On the other hand, the impact on the quality of the meat, whether in terms of its nutritional or organoleptic characteristics, seems to be limited or at least more difficult to establish because it depends on many factors related to the stage of animal rearing, the algal species used and even its mode of administration.

Despite these reservations, the use of algae in pig farming has many advantages. Among these, the beneficial effects on animal health related to the digestive microbiota appear to be essential. The presence of polysaccharides such as laminarin in some algae, with well-established antibacterial properties, is a major asset for limiting the use of antibiotics in pig farming.

Such an opportunity to valorize seaweed in pig farming could represent an alternative strategy for a sector marked by an industrial type of intensive production. In France, it would be all the more relevant as Brittany, the first pork production region, is also the first algae production region.

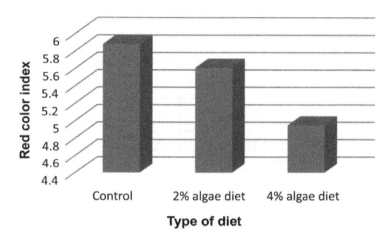

Figure 4.17. *Effect of providing the brown alga* Macrocystis pyrifera *as a dietary supplement on the color intensity of meat of fattening pigs (from Jerez-Timaure et al. 2021)*

4.3. Horse nutrition

The use of seaweed in horse nutrition is part of traditional practices in Europe and especially in the northern regions of the continent (see Chapter 1) (Delaney et al. 2016; Fleurence 2021b).

Commercial seaweed-based products are currently available for the diet of horses, foals and ponies. For this type of application, the company Ocean Harvest Technology[2] offers a specific product called OceanFeed Equine. This product is made from algae from cold European waters and warm South Asian waters, according to the company's brochure.

The composition of the product in terms of seaweed species is an industrial secret, so it is difficult to know what it is. However, as the company is based in Galway (Ireland), it is likely that brown algae such as Fucales (*Ascophyllum nodosum*, *Fucus* sp.), which are common on these coasts, are used in the product.

---

2 https://oceanharvesttechnology.com/.

Taking into account the high content of iodine in *Laminaria* and the risks of intoxication by this element, Laminariales are generally little used for the feed of horses (Zafren 1935). As such, if they are not de-iodized, it seems unlikely that they contribute to the elaboration of the commercialized product.

On the other hand, the integration of green algae of the genus *Ulva* appears possible with regard to the availability of its biomass on the Irish coasts.

The product is formulated as pellets, flour or flakes. Its approximate composition is given in Table 4.8.

Compound	% of gross product
Humidity	14
Fats	0.5
Ashes	30
Protein	10
Carbohydrates	45
Fibers	5
ADF (crude cellulose equivalent)	18

Table 4.8. *Chemical composition of OceanFeed Equine (from Ocean Harvest Technology)*

The administration of the product differs according to the type of horse and its stage of development (see Table 4.9).

Type of animal	Gram/day/animal
Weaned foal	15
One-year-old foal	30
Ponies	30
Adult horse	45

Table 4.9. *Conditions of administration of OceanFeed Equine product by stage of equine development (from Ocean Harvest Technology)*

The action on digestive health, thanks to the presence of prebiotics, is the main nutritional claim made by the manufacturer. The effect of the product on the improvement of the shine of the horse's coat is also put forward by the latter. This aesthetic aspect, which is very important during horse shows, is particularly highlighted by the manufacturer on social media[3]. The richness in mineral elements of the seaweed is also part of the nutritional asset developed during the commercial argumentation.

A few studies, limited in number compared to those carried out on ruminants and pigs, describe the impact of the addition of algae in the diet of equids. A study conducted on mature stallions revealed that the addition of an extract of *A. nodosum* at a level of 1.5% of the feed ration had no effect on feed intake, weight gain of the animal or semen quality (García-Vaquero 2019).

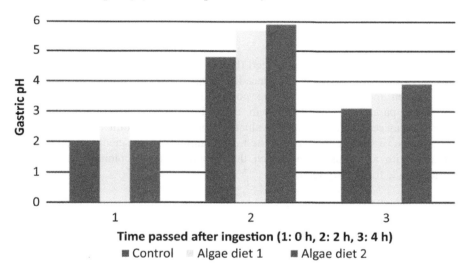

Figure 4.18. *Effect of supplying a mineral supplement (Calmin) based on red algae on the gastric pH of neutered horses and mares (from Jacobs et al. 2020). For a color version of this figure, see www.iste.co.uk/fleurence/algae.zip*

Another more recent study shows that the addition of the red calcareous algae *Litho thamnion corallioides* and *Phymatolithon calcareum* to horse feed has a positive effect on regulating gastric pH, thus limiting the risk of ulcers (Jacob et al. 2020). The algae are supplied in the form of a feed supplement called Calmin to geldings and mares. The animals are fed a ration composed of a concentrated feed (Purina Omolene 100) (2 kg/day) and a hay supplement (1.5% of the ration/day).

---

3 https://www.facebook.com/oceanfeedequine.

Calmin is administered at the unitary concentration recommended by the supplier (alga diet 1) or at twice this concentration (alga diet 2). As shown in Figure 4.18, the intake of the calcareous algae concentrate significantly buffers the pH of the equine gastric juice especially two hours after feeding. This buffering effect, which limits gastric acidity, is, however, weaker four hours after ingestion of the feed ration.

Although algae have been used for centuries in horse diets (see Chapter 1), they have been little studied as functional feeds in equids. Indeed, research efforts have mainly focused on the nutrition of ruminants and some monogastrics such as pigs. This observation is probably linked to the particular status of equids in animal production. Although horses are domestic animals, they no longer meet the status of livestock animals that they had in the past. Their breeding is now more associated with recreational activities than with the supply of meat for human consumption.

4.4. Poultry nutrition

The addition of algae to poultry feed (laying hens, chickens, ducks, turkey) is a relatively widespread practice. For decades, it has been the subject of numerous studies, given the importance of the poultry industry. These studies focus on zootechnical parameters such as growth, as well as on meat and egg quality. The impact of seaweed feeding is also evaluated from a health point of view, i.e. the reinforcement of the health status of the treated animals. This scientific approach is correlated with an industrial approach that proposes many commercial products including algae for poultry feed.

The effect of incorporating seaweed in the diet of laying hens and the impact on egg quality is particularly well documented. As early as 1960, Hoie and Sannan (Jensen 1963) observed a slight darkening of the egg yolk when 5% seaweed meal was added to the diet of laying hens. The addition of brown seaweed meal (*Ascophyllum nodosum, Fucus vesiculosus, Fucus serratus*) to the diet of laying hens has an effect on the carotenoid content of the egg yolk and thus on the intensity of its coloration (see Table 4.10).

At *A. nodosum* levels of 15% and above, some animals produce eggs with a distinct orange yolk color. This suggests that not all animals have the same ability to transfer algal carotenoids into the yolk. Eggs with a distinct orange color are characterized by significantly higher lutein and zeaxanthin levels than those observed in other eggs (see Table 4.11).

Carotenoids	Control diet	Diet 10% A. nodosum	Diet 15% A. nodosum
Carotene + cryptoxanthin	147	157	173
Lutein	648	683	520
Zeaxanthin	342	362	264
Related fucoxanthin	0	Traces	194

Table 4.10. *Effect of* Ascophyllum nodosum *on the carotenoid content (μg/100 g fresh yolk) of egg yolk of laying hens (from Jensen 1963)*

Carotenoids	Egg yolk	Orange egg yolk
Carotene + cryptoxanthin	156	208
Lutein	300	960
Zeaxanthin	139	515
Fucoxanthin and related compounds	168	246
Total	763	1929

Table 4.11. *Effect of a 15%* Ascophyllum nodosum *diet on the carotenoid content (μg/100 g fresh yolk) of the yolk of yellow or orange eggs of laying hens (from Jensen 1963)*

A more recent study (Strand et al. 1998) shows that feeding *Fucus* sp. meal to 15% of the diet of laying hens significantly improves the yolk color of the eggs. The yolk color, as measured by the Roche scale, is on average four times more intense than that reported for eggs from hens not fed the algal supplement (see Figure 4.19).

The dietary acceptability of seaweed meal by laying hens is still a limit to this approach to improving egg yolk color. An old study (Vale and Smetana 1965) shows that depending on the species of seaweed used and their incorporation rate, the impact on egg color is low or even null.

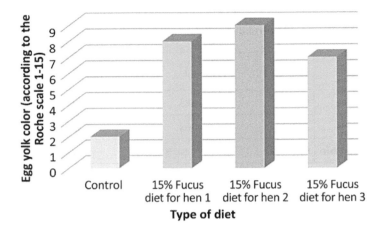

Figure 4.19. *Effect of a 15% Fucus sp. diet on egg yolk color according to the Roche scale (1–15) (from Strand et al. 1998)*

Indeed, an experiment based on the use of brown algae (*Ecklonia radiata, Sargassum* sp.) or green algae (*Ulva lactuca*) in the feeding of pullets leads to disappointing results in this field. The palatability of the brown algae concerned is low and their consumption in small quantities does not allow us to observe a change in the intensity of the egg yolk compared to that of the eggs produced by hens on the control diet (without algae). In contrast, the palatability for ulva meal is sufficient for the animals to consume different amounts of meal (2–12% of the feed ration). A content of 12% of algal meal is, however, necessary to achieve a yolk coloration that meets the color standards established for the Australian market, the country in which this experiment is being conducted.

This study therefore highlights the importance of the choice of algal species in the nutrition of laying hens in an approach aiming to improve yolk color.

The impact of dietary intake of red algae (*Chondrus crispus, Sarcodiotheca gaudichaudii*) on the egg quality of laying hens has also been studied (Kulshreshtha et al. 2014). The algal supplement is incorporated as a powder at 0.5, 1 and 2% in the feed ration. An effect on egg and yolk mass is observed for the 1% concentration of algae in the diet (see Table 4.12). For the other concentrations, no significant effect is reported for these weight parameters illustrative of egg quality.

Other qualities such as the thickness of the egg white (albumin), the color of the yolk or the thickness of the shell do not seem to be affected by the diet supplemented with red algae.

These results confirm that the impact of an algae-based diet on egg production remains highly dependent on the species used and their concentration of use in the final ration.

Diet	Mass of the egg (g)	Mass of fresh egg yolk (g)
Control	63.43	17.27
CC 1%	66.82	18.48
SG 1	65.63	18.72

Table 4.12. *Effect of the addition of red algae flour (*Chondrus crispus *or CC,* Sarcodiotheca gaudichaudii *or SG) on egg and yolk weight (from Kulshreshtha et al. 2014)*

The effects of supplementing the diet of laying hens with algae on the nutritional quality of the egg are also known. The main impact reported is the iodine enrichment of the product (Kaufman et al. 1998; Autret and Madec 2003).

An experiment with six-month-old laying hens fed a red algal meal (*Eucheuma spinosum*) at 5 or 10% of the feed ration showed a positive effect on the iodine content of the egg yolk and egg white (Kaufman et al. 1998) (see Figures 4.20 and 4.21).

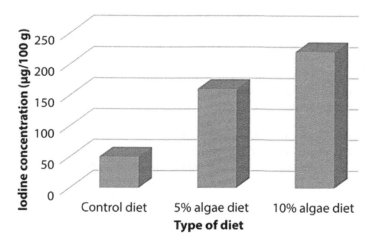

Figure 4.20. *Effect of feeding red algal meal (*Eucheuma spinosum*) on the iodine content of egg yolk after four weeks of feeding (from Kaufman et al. 1998)*

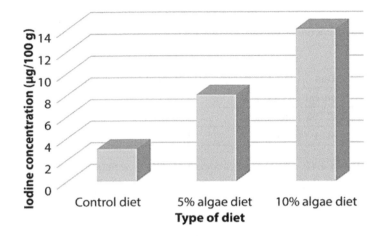

Figure 4.21. *Effect of feeding red algal meal (*Eucheuma spinosum*) on the iodine content of egg white after four weeks of feeding (from Kaufman et al. 1998)*

In terms of valorization, a patent application concerning the iodine enrichment of eggs from hens fed with a mixture of algae with an iodine content of at least 1,000 mg/kg was published in 2003 (Autret and Madec 2003). The species used are *Ascophyllum nodosum* and *Lessonia nigrescens* at 2.5% of the feed ration.

As previously mentioned (see sections 4.1 and 4.2), the use of seaweed in the diet of livestock (cattle, pigs, poultry) generates iodine enrichment of the products from their rearing (milk, meat, eggs). This approach appears to be a complementary strategy to the use of iodized culinary salt in compensating for the iodine deficiency of continental populations.

In addition to algal meals, the effect of dietary intake of algal polysaccharides on egg quality has also been studied recently (Guo et al. 2020). In this experiment, polysaccharides extracted from the genus *Enteromorpha* were provided to 16-month-old laying hens. Three diets integrating these at concentrations of 1,000, 2,500 and 5,000 mg/kg of the feed ration were administered to the animals for six weeks. The addition of polysaccharides resulted in a significant improvement in yolk color (+25% for the 2,500 mg/kg dose). Shell thickness was also increased following the algal polysaccharide diet (see Figure 4.22).

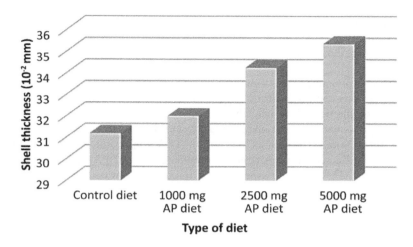

Figure 4.22. *Impact after six weeks of algal polysaccharide (AP) administration on eggshell thickness of white Lohmann laying hens (from Guo et al. 2020)*

Shell resistance to breaking is also markedly enhanced following ingestion by hens of *Enteromorpha* polysaccharides (see Figure 4.23).

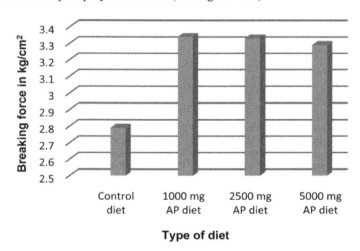

Figure 4.23. *Impact after six weeks of algal polysaccharide (AP) administration on eggshell fracture resistance of white Lohmann laying hens (from Guo et al. 2020)*

Being part of the Ulvales, *Enteromorpha* sp. is a ubiquitous and easily accessible alga. By focusing on green algae, a species less valued than brown or red algae, this type of work highlights the interest of using Ulvales in poultry production.

The interest in using algae in poultry feed is not limited to laying hens and egg production. It also applies to broiler nutrition. In both cases, the provision of algae as a dietary supplement allows for the reduction of microbial load in the digestive tract, the improvement of meat or egg quality, and the enhancement of the immune status of the animals (Morais et al. 2020). This improvement of the immune mechanism is reflected in an increase in immunoglobulin (IgA, IgG) levels in the animals' plasma. This inducing aspect of the immune system will be developed later (see section 5.3).

The effect of dietary supplements on poultry health is now well documented (Kulshreshtha et al. 2020). Extracts of brown, red or green algae are described in the scientific literature as having antibacterial and even antiviral properties (see Table 4.13).

Algae	Rate of inclusion in the food ration	Antimicrobial responses
Laminaria japonica + cecropine	– Seaweed (1, 3, 5%) – Cecropine (0.03%)	– Antibacterial and antiviral properties – Significant decrease in *E. coli* levels
Ascophyllum nodosum	0.1%	– Antibacterial activities – Reduction of *Campylobacter jejuni* contamination in young chickens
Chondrus crispus	2%	– Antibacterial activities
Sarcodiotheca gaudichaudii	4%	– Reduction of the negative effects of the *Salmonella enteritidis* agent on egg production
– *Grateloupia filicina* – *Ulva pertusa* – *Sargassum qingdaoense*	20–500 mg/mL of sulfated polysaccharides	– Antiviral activities – Inhibition of the avian influenza virus
Ulva clathrata + **fucoidan**	0.1–1,000 µg/mL	– Antiviral activities – Inhibition of Newcastle disease (paramyxovirus type 1)

Table 4.13. *Examples of extracts of brown, green or red algae and algal polysaccharides with activity against poultry (laying hens or broilers) pathogens (from Kulshreshtha et al. 2020)*

More generally, at least some species of algae seem to have an effect on reducing the mortality rate of poultry. This is reported for brown algae belonging to the species *Undaria pinnatifida* or *Hizikia fusiforme* and also for green algae belonging to the *Ulva rigida* species or red algae such as *Palmaria palmata* (Michalak and Mahrose 2020) (see Table 4.14). Apart from the species effect, it should be noted that the use of fermented algae produces better results in terms of reduction of the mortality rate at the farm level (– 9 times). Fermentation thus appears to be a useful process that probably improves the production of prebiotics that are essential to the microbiota of the animals and thus to their general health.

Algal species	Type of poultry	Decrease in mortality rate compared to the control group
Undaria pinnatifida **(0.5%)** **Fermented** *U. pinnatifida*	Broiler chickens	– 4.5 times (algae) – 9 times (fermented algae)
Hizikia fusiforme **(0.5%)** *Hizikia fusiforme* **(fermented)**	Broiler chickens	– 3 times (algae) – 9 times (fermented algae)
Palmaria palmata **(2.4%, 3%)**	Broiler chickens	– 67% (2.4%) – 67% (3%)
Ulva rigida **(2%, 4%, 6%)**	Broiler chickens	– 4 times (2%) – 4 times (4%) – 17% (6%)

Table 4.14. *Effect of feeding seaweed on the decreased mortality rate of broilers (from Michalak and Mahrose 2020)*

In addition to laying hens and broilers, the impact of an algal diet on other poultry such as ducks and turkeys is well documented. The dietary contribution of the red alga of the genus *Polysiphonia* has been shown to be beneficial for certain quality aspects of carcasses (El-Deek and Brikaa 2009). This is particularly interesting as *Polysiphonia* algae are widespread on the Mediterranean and Atlantic coasts (see Figure 4.24) and constitute a readily accessible biomass.

The study was conducted on the feeding of ducks at the beginning of rearing (one-day-old ducklings) and on animals entering the last phase of rearing (36 days old). The animals were fed ad libitum with a diet containing 0–15% algae in the feed ration. In the early rearing phase, algal intake, which could be as high as 12% of the diet, did not have a significant effect on feed intake or feed conversion rate by the animals. This was true regardless of the form of algae administration (pellets or pureed).

Figure 4.24. *Red algae of the genus* Polysiphonia *in the center of a basin on the foreshore of the Atlantic coast (photo credit © J. Fleurence, 2021). For a color version of this figure, see www.iste.co.uk/fleurence/algae.zip*

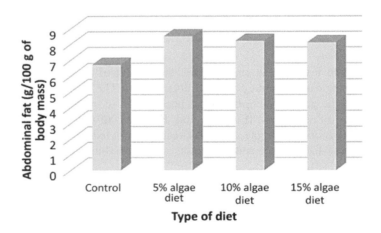

Figure 4.25. *Effect of incorporating the alga* Polysiphonia *spp. (pellets or pureed) in the diet of ducks at the end of rearing on the abdominal fat content of the animals (from El-Deek and Brikaa 2009)*

In animals aged 36–65 days (end of rearing period), the addition of algae had no impact on growth and feed intake. Adult animals could tolerate 15% algae in the feed ration, which was not the case for ducklings (12% maximum). On the other hand, the addition of algae in the diet had an influence on the carcass quality of the animals at the end of the rearing period. Indeed, the contribution of 5–10% of algae in the ration of ducks in the final rearing period (aged 35–65 days) led to a significant increase in the weight of the chest muscles (duck breasts) and the abdominal fat content of the animals (see Figure 4.25). An improvement in thigh texture was most clearly observed with the 5 and 10% seaweed diets (see Figure 4.26). The 15% diet mainly resulted in a gain in the texture of the breast. Regarding other organoleptic qualities (e.g. aroma, taste, juiciness and color) of the meat, no significant differences were reported depending on the incorporation rate of seaweed in the diet at the end of rearing.

This study shows that algae, at least *Polysiphonia*, can be used at the beginning or end of duck rearing without a negative impact on animal growth and on the main quality criteria associated with the carcass.

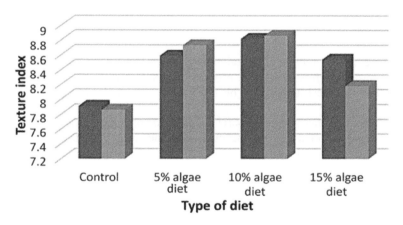

Figure 4.26. *Effect of incorporating* Polysiphonia *spp. seaweed (pellets or pureed) in the diet of ducks at the end of rearing on the breast and leg texture of the animals (from El-Deek and Brikaa 2009). For a color version of this figure, see www.iste.co.uk/ fleurence/algae.zip*

Another study on the use of the brown alga *Laminaria japonica* in duck diets supports the previous results (Islam et al. 2014). Indeed, the addition of algae in the diet of the animals has no effect on their growth as it had previously. Furthermore, the composition of the meat is little modified by the algal diet, except for the fat content, which is significantly higher for animals fed a diet incorporating 1% algae (see Table 4.15). This finding is consistent with the findings for the *Polysiphonia* diets (see Figure 4.25).

	Control diet	Diet 0.1% algae	Diet 0.5% algae	Diet 1% algae
Moisture (g)	76.7	77.1	76.8	76.5
Ash (g)	1.67	1.86	2.65	1.30
Fat (g)	0.51	0.32	0.50	0.63
Protein (g)	20.8	20.1	20.7	20.8

Table 4.15. *Effect of the dietary intake of the alga* Laminaria japonica *on the chemical composition of duck meat (per 100 g) (from Islam et al. 2014)*

4.5. Nutrition of rabbits

The breeding of rabbits for food purposes, or cuniculture, remains marginal compared to other terrestrial animal productions. Nevertheless, the impact of algae on the nutrition of these animals has been the subject of some studies.

The effects of a diet with Laminaria (*Laminaria* spp.) as its basis on the growth of New Zealand rabbits and on the quality of the meat produced are now well documented (Rossi et al. 2020). In this experiment, 150 animals were fed three different diets (no algae, 0.3% algae, 0.5% algae) over a period of 42 days. The effects of the algal diets on animal growth were particularly noticeable after 42 days of feeding. The diet with 0.3% algae was particularly effective in increasing the weight of the animals (see Figure 4.27).

Algae-based diets also had an impact on the nutritional quality of the meat. A significant decrease in muscle cholesterol content was observed with the 0.3% seaweed diet (see Figure 4.28). In contrast, the diet with 0.5% seaweed appeared to be slightly less effective in reducing the cholesterol content of the meat produced. The levels of natural antioxidants vitamins A and E were also impacted by the addition of seaweed to the diet. The 3% diet significantly increased the levels of vitamins A and E in semimembranosus muscle by 110% and 38.5% respectively.

Algae-based diets also had an effect on the texture of the thigh. They significantly improved the tenderness and juiciness of the product while decreasing its stringiness (see Figure 4.29). The impact on these three organoleptic characteristics was particularly noticeable for the diet with 0.3% Laminaria.

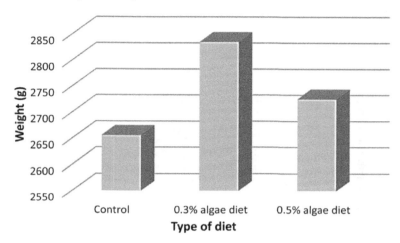

Figure 4.27. *Impact of algal (*Laminaria spp.*) diets on animal weight after 42 days of feeding (from Rossi et al. 2020)*

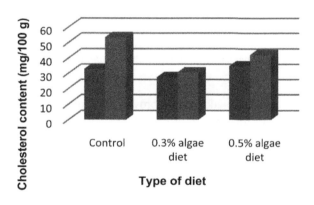

■ Thoracic longissimus muscle and lumborum

■ Semimembranosus muscle

Figure 4.28. *Effect of diets with* Laminaria *algae (*Laminaria spp.*) as their basis on the cholesterol content of rabbit meat (from Rossi et al. 2020). For a color version of this figure, see www.iste.co.uk/fleurence/algae.zip*

Apart from these aspects, the addition of seaweed to the diet seemed to have an effect on the quality of the meat during storage at 4°C. Both algae-based diets had better resistance of the meat to lipid oxidation, i.e., rancidity. Indeed, after 72 hours of conservation, the meat from animals fed with the two algae-based diets showed a 28% lower oxidation rate than that recorded for the control meat. This phenomenon is also known, as previously mentioned, for pigs (see section 4.2).

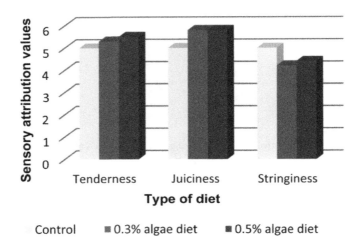

Figure 4.29. *Effect of diets based on* Laminaria *algae (*Laminaria spp.*) on the texture of rabbit thigh (from Rossi et al. 2020). For a color version of this figure, see www.iste.co.uk/fleurence/algae.zip*

This study shows that the addition of brown algae of the genus *Laminaria* in small quantities (0.3% of the feed ration) has a positive effect on the growth of the animals as well as on certain aspects of the quality of the meat: the content of cholesterol, vitamins and certain sensory characteristics of major interest to the consumer (texture, juiciness, stringiness).

Macroalgae are inputs traditionally used or tested in the nutrition of land-based livestock (sheep, cattle, equines, pigs, poultry and rabbits). The polysaccharides present in brown algae (alginates, laminarine, fucoidans) often carry prebiotic activities that have an influence on the biodiversity of the intestinal microbiota and thus on the digestive health of animals.

Some microalgae (*Chlorella* sp., *Haematococcus* sp., *Scenedesmus* sp.) or cyanobacteria (*Arthrospira* sp.) can also be used as feed supplements in terrestrial animal production.

Arthrospira sp., better known as spirulina, is used in chicken feed to intensify the orange coloration of the flesh. This effect is probably due to the presence of a large diversity of carotenoid pigments (β-carotene, xanthophylls, zeaxanthin and canthaxanthin) in this cyanobacterium (Fleurence 2021a).

Similarly, supplying laying hens with the microalga *Nannochloropsis gaditana* at 5% of the feed ration results in bright orange yolks (Fleurence 2021a). This finding reveals the existence of an efficient transfer of the carotenoid pigments contained in the alga during the feeding of the animal.

The contribution of the microalga *Scenedesmus obliquus* in calf feeding is mainly described as improving the role of the gut in the digestive process.

Despite this, the impact of the introduction of microalgae into the diet of terrestrial animals is less documented and of more recent study than that reported for macroalgae.

This observation will appear somewhat different from the nutrition of animals produced by aquaculture (fish, mollusks, crustaceans).

4.6. Nutrition of animals produced by aquaculture

In 2018, global fisheries and aquaculture production was estimated at 178 million tons (live weight) (FAO 2020). Aquaculture accounted for 46% of the tonnage produced (see Figure 4.30). As such, it was close to the level of fishing, which was estimated at 54% of the tonnages taken from the environment. The species harvested belonged mainly to three main groups, namely fish, crustaceans and mollusks. The scarcity of wild sea urchin resources has also recently allowed the development of an aquaculture dedicated to echinoderms or echinoculture.

Algae are often used as food for these aquatic animals. Microalgae and cyanobacteria are part of the diet of many mollusks and crustaceans. They can also be used indirectly in the diet of fish. Macroalgae are more often reserved for the nutrition of fish and certain gastropod mollusks such as abalone.

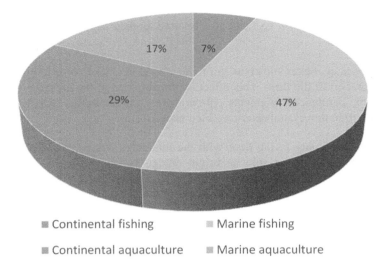

Figure 4.30. *Distribution of world aquaculture and fisheries production (wet tonnage) (from FAO 2020). For a color version of this figure, see www.iste.co.uk/fleurence/algae.zip*

4.6.1. *Fish*

The use of algae in the diet of aquacultured fish has been the subject of numerous studies. The impact of this feeding method on the growth of the animals, their health and their bodily composition is often well documented. The fish in these kinds of studies are freshwater, diadromous or marine species. The algae tested are mainly macroalgae belonging to the three main groups Phaeophyta (brown algae), Rhodophyta (red algae) and Chlorophyta (green algae). The algae are, in most cases, included in the feed ration in the form of meal and at extremely variable levels depending on the species of algae and fish (see Table 4.16).

The effects on animal growth vary according to many parameters such as the diet (herbivore, carnivore, omnivore), the algal species used and its rate of incorporation in the ration. This can be observed with red sea bream for which the administration of *Ascophyllum nodosum* or *Undaria pinnatifida* algae at the 5% level does not produce the same effects on animal growth (Wan et al. 2019). On the same fish species, feeding 3% or 5% *Porphyra yezoensis* in the diet has a positive effect on growth parameters (increase in final animal weight and daily growth rate; Mustafa et al. 1995).

Algal species	Fish	Diet	Inclusion rate of algae in the diet	Effect on growth
Ulva lactuca (green alga)	Sea bream	Omnivore	2.6–7.8% 14.6–29.1%	No effect on final weight and decrease in growth parameters
Ulva lactuca (green alga)	Seabass	Carnivore	5–10–15%	Increase in final weight and specific growth rate*
Ulva rigida (green alga)	Sea bream	Omnivore	25%	Increase in final weight and specific growth rate*
Ulva rigida (green alga)	Seabass	Carnivore	5–10%	Final weight loss with increased inclusion rate
Ulva rigida (green alga)	Nile tilapia	Herbivore	10–20–30%	Final weight loss and decrease in relative growth rate to 30%
Gracilaria bursa-pastoris (red alga)	Seabass	Carnivore	5–10%	No impact on final weight and daily growth gain

Algal species	Fish	Diet	Rate of inclusion of algae in the diet	Effect on growth
Gracilaria pygmaea (red alga)	Rainbow trout	Carnivore	3–6–9–12%	Increase in final weight to 3, 6 and 9% and decrease in specific growth rate* from 12%
Palmaria palmata (red alga)	Atlantic salmon	Carnivore	5–10–15%	No impact on final weight and growth gain
Ascophyllum nodosum (brown alga)	Red sea bream	Carnivore	5–10%	Reduction of the final weight
Undaria pinnatifida (brown alga)	Red sea bream	Carnivore	5–10%	Increase in final weight for the 5% concentration
Porphyra yezoensis (red alga)	Red sea bream	Carnivore	3–5%	Increase in final weight and daily growth rate

* Specific growth rate: (final weight − initial weight/number of days of rearing) × 100)]

Table 4.16. *Some examples of the impact of algal use on the growth of freshwater, diadromous and marine fish (from Mustafa et al. 1995 and Wan et al. 2019)*

Sea bass (*Dicentrarchus labrax*) is a marine species at the origin of the development of aquaculture in Southern Europe (Greece, Italy, France, Spain). The impact of the contribution of algae in the nutrition of juveniles has been the subject of particular interest in view of the economic interest of this sector.

In particular, the effects of diets based on Gracilaria (*Gracilaria* spp.), ulva (*Ulva* spp.) and Fucus (*Fucus* spp.) on the diet of juveniles are well documented in a 2016 study (Peixoto et al. 2016). The algae provided were integrated up to 2.5% or 7.5% into the diet. In addition, a mixture of the three algae administered at 2.5% of the final ration is also a part of the tested diets. The addition of the algae to the diet of juveniles did not induce any effect on the growth of the animals, on their feed intake capacity, on their feed conversion rate or on the protein efficiency coefficient. Thus, the main zootechnical parameters seem to be little affected by the inclusion of algae in the diet of juveniles.

On the other hand, biochemical criteria and more particularly certain enzymatic activities are more or less affected by the presence of algae in the diets. The digestive enzymes (trypsin, chymotrypsin) present an activity that is little impacted according to the nature of the diets. This is not the case for the lipase activity which is strongly increased in the presence of Gracilaria or Fucus extracts (see Figure 4.31). On the other hand, the dietary intake of ulva had no effect on lipase activity compared to the control diet.

At the hepatic level, a significant increase in glutathione peroxidase activity was observed (see Figure 4.32), whatever the algal species used. However, the latter is more important with the Gracilaria extracts or the algal mixture. This increase accentuates the resistance of the animal to oxidative stress linked to the generation of peroxide ions.

In addition to the aspects previously described, the addition of algae in the nutrition of juvenile sea bass shows an effect on the immune status of treated animals. The inclusion of algae at 7.5% of the diet has a negative influence on the complement pathway which is considered an innate immune response. This negative effect is not observed, however, for diets incorporating lower levels of algae (2.5%).

Algal diets also have an effect on the lysosomal activity of treated animals. In particular, lysosomal activity is significantly increased in the presence of ulva, Gracilaria or Fucus at levels of 2.5% of the diet ration (see Figure 4.34).

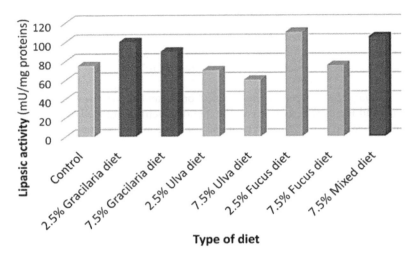

Figure 4.31. *Effect of algal-based diets on the intestinal lipase activity of juvenile sea bass (*Dicentrarchus labrax*) (from Peixoto et al. 2016)*

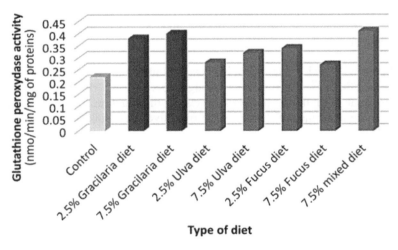

Figure 4.32. *Effect of algae-based diets on the activity of hepatic glutathione peroxidase of juvenile sea bass (*Dicentrarchus labrax*) (from Peixoto et al. 2016)*

The work carried out shows that the addition of seaweed to the feed of juvenile sea bass does not have a negative impact on the growth and development of the animals. Some effects are, however, reported on the antioxidant and immune status of the animals. A beneficial effect on the complement pathway and lysosomal activity is found with an ulva-based diet (2.5%) (see Figures 4.33 and 4.34).

Feeding of Livestock 109

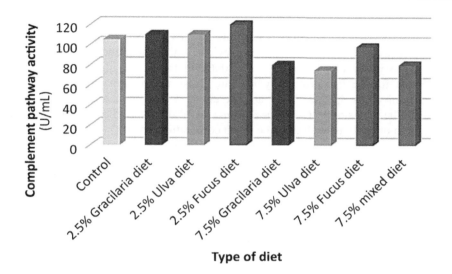

Figure 4.33. Effect of algal-based diets on the complement pathway in juvenile sea bass (Dicentrarchus labrax) (from Peixoto et al. 2016)

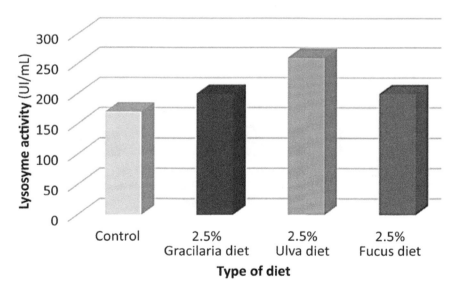

Figure 4.34. Effect of algal-based diets on lysosomal activity in juvenile sea bass (Dicentrarchus labrax) (from Peixoto et al. 2016)

A similar study involving the use of two species of Gracilaria (*Gracilaria bursa-pastoris, Gracilaria cornea*) and one species of ulva (*Ulva rigida*) in the nutrition of juvenile sea bass shows, as previously, the importance of the inclusion rate of algae in the diet (Valente et al. 2006). In this study, algae are used as a replacement for the fish protein hydrolysate classically present in the diet of sea bass, a carnivorous species. They are present at 5–10% in the diet provided to juveniles. The inclusion of *Gracilaria cornea* at 10% reduces the final growth gain compared to other diets, including the one incorporating 5% of *G. cornea*. This level of inclusion also has a negative effect on daily weight gain (see Figure 4.35). The 10% GC diet also shows a negative effect on the digestibility of the feed provided to juveniles (see Figure 4.36). Apart from this, the administration of *G. bursa-pastoris* was found to have a beneficial impact on daily weight gain as well as on the final weight of the animal, regardless of the integration rate of the algal ingredient (see Figure 4.35). This last result suggests that this algal species is an ingredient that can be substituted for fish hydrolysate during the rearing of juveniles.

This study highlights the differentiated impact produced by the contribution of algae depending on the species used, as well as on the rate of incorporation in the final ration. As before, positive effects are often observed for relatively low levels (2.5–5%).

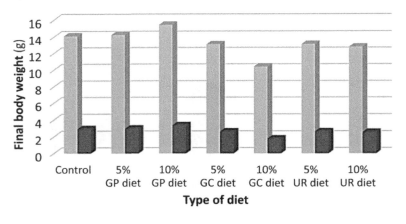

Figure 4.35. *Impact of the use of* Gracilaria bursa-pastoris *(GP),* Gracilaria cornea *(GC) and* Ulva rigida *(UR) species on body weight gain of juvenile sea bass* (Dicentrarchus labrax) *(from Valente et al. 2006)*

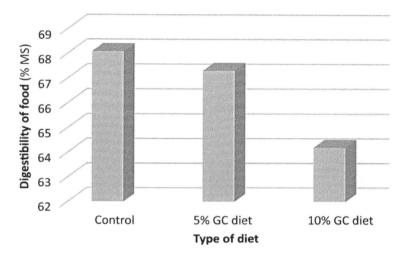

Figure 4.36. *Impact of using different percentages of the species* Gracilaria cornea *(GC) on the digestibility of the feed provided to juvenile sea bass (*Dicentrarchus labrax*) (from Valente et al. 2006)*

The impact of algal supplementation in the diet of juvenile salmon (*Salmo salar*) on their growth and lipid metabolism has also been studied (Norambuena et al. 2015). This study was based on the use of two commercially available products, one with macroalgae (Verdemin) as its basis and the other based on microalgae (Rosamin). This approach differed from the previous ones, which mainly involved the use of macroalgae in powder or meal form in the fish diet. The product Verdemin was obtained from the macroalga *Ulva ohnoi* and the product Rosamin from the diatom *Entomoneis* spp.

The products were included in isolation, at 2.5 and 5% of the diet, or in combination. As before, the inclusion of algae in the diet did not have any positive or negative effect on the growth of the animals. On the other hand, the Rosamin-rich diet (5%) induced a significant increase in the content of long-chain polyunsaturated fatty acids of the omega-3 series. This was particularly noticeable in the fillets obtained from salmon which were fed this diet (see Figure 4.37). This result is interesting because it shows that the supply of algae, and more particularly microalgae, can have a positive effect on the quality of salmon flesh, a product valued for its content of unsaturated fatty acids of the omega-3 and omega-6 series (eicosapentaenoic acid). Another result of this study is the impact of the algal ingredients consumed on the health of the aquatic livestock. Indeed, animals fed the diet incorporating the Verdemin product did not show any deaths, which does not seem to be the case for fish fed with the other diets (see Figure 4.38). This finding is likely related to the presence of ulva in Verdemin, as this species has previously been described to improve stress response and disease resistance in sea bream and salmon (Fleurence et al. 2012).

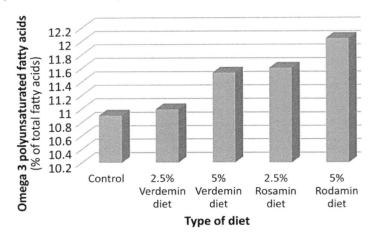

Figure 4.37. *Effect of Verdemin (*Ulva ohnoi*) and Rosamin (*Entomoneis spp.*) diets on the content of long-chain polyunsaturated fatty acids in salmon (*Salmo salar*) fillets (from Norambuena et al. 2015)*

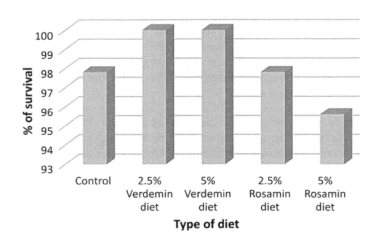

Figure 4.38. *Effect of diets based on Verdemin (*Ulva ohnoi*) and Rosamin (*Entomoneis spp.*) on the survival rate of juvenile salmon (*Salmo salar*) (from Norambuena et al. 2015)*

Another study using commercial product AquaArom, containing an extract of *Laminaria* sp., gave more encouraging results in the growth of smolts (young salmon still having the morphological characteristics of juveniles) (Kamunde et al. 2019). The use of the product at the level of 3% of the diet significantly increased the daily weight gain of animals on this diet (see Figure 4.39). Compared to the control diet, the 3 and 10% diets had a positive impact on the size and final weight of the animals. In contrast, the 6% diet showed a neutral effect on these zootechnical parameters. The algal diets (3 and 10%) also increased the protein efficiency ratio (dietary protein supplied/protein produced) and improved the feed conversion process. This last aspect is particularly important in the development of a more feed-input efficient and less nitrogenous waste-generating aquaculture system.

In addition to these aspects, algal feed increased the antioxidant activity of plasma in animals fed this type of diet. This was reflected in an increase in glutathione concentration (see section 5.3.1) and in superoxide dismutase and catalase activities (see Figure 4.40). The contribution of algae in the nutrition of young salmon also appears to have an impact on mitochondrial respiration by limiting the effects of a rapid rise in temperature on the latter. This last point suggests that the contribution of algae could moderate the impact of certain stress factors such as an acute increase in temperature.

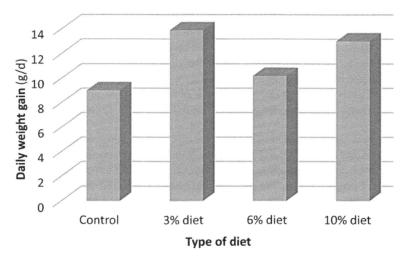

Figure 4.39. *Effect of AquaArom (*Laminaria *sp.) diets on the daily weight gain of smolts (from Kamunde et al. 2019)*

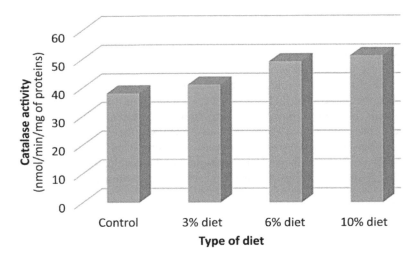

Figure 4.40. *Effect of AquaArom (*Laminaria sp.*)-based diets on plasma catalase activity of smolts (from Kamunde et al. 2019)*

Studies carried out on carnivorous fish species (sea bass, salmon) or omnivores (sea bream) show that the addition of seaweed to the diet of these animals does not have a negative effect on their growth and can be a substitute in some cases for the addition of fishmeal to the diet. It is often associated with reduced mortality in farms, mainly due to an increase in the antioxidant capacity of the animals coupled with an increase in immune status (see section 5.3.1). The integration of algae in the nutrition of carnivorous fish appears to be a possible way to promote a sustainable aquaculture that no longer relies on industrial fishing for its development.

Tilapia is a herbivorous freshwater fish and had an aquaculture production of nearly 4.2 million tons in 2016 (FAO 2018). Nile tilapia (*Oreochromis niloticus*) is a major source of animal protein in Africa and Asia. The impact of providing algae or cyanobacteria (spirulina) on the growth and health of reared animals is now well documented. The supply of a daily dose of 29 g of spirulina (expressed as dry matter) per kilogram of body weight results in a daily growth of 14.4 g/kg in both blue tilapia (*Tilapia aurea*) and the plant-eating buffalo fish (*Ictiobus cyprinellus*) (Fleurence 2021a). The inclusion of the cyanobacterium *Spirulina* in the tilapia diet also has a health benefit for the reared stock. In the face of infection by the *Aeromonas hydrophila* bacterium, tilapia fed a diet containing 10 g of spirulina per kilogram of feed had a farm mortality rate of just under 10%. In the absence of spirulina, this rate is 80% only four days after contact with the pathogen (Fleurence 2018).

The effects of including macroalgae in the diet of tilapia and specifically Nile tilapia (*Oreochromis niloticus*) have been the subject of a recent study (Younis et al. 2018). The purpose of this experiment was to replace fishmeal with different levels of *Gracilaria arcuata*, a red alga belonging to a protein-rich genus (expressed up to 30% on a dry matter basis). The algae were incorporated at levels of 20, 40 or 60% of the feed ration. The incorporation of seaweed into the tilapia diet had a negative effect on growth, whether expressed as final weight gain or specific growth rate (see Figure 4.41). This negative impact was more pronounced the higher the level of alga integration into the diet. This having been said, the incorporation of algae had no effect on the survival rate of the animals, which remained at 100% regardless of the diet applied. The best feed conversion rate (feed intake in g/weight gain in g), which corresponded to the lowest ratio value, was found to be for the control diet. The highest protein efficiency coefficient, which measures the efficiency of assimilation of dietary protein and its conversion into animal protein, was also found to be for the control diet (see Figure 4.42).

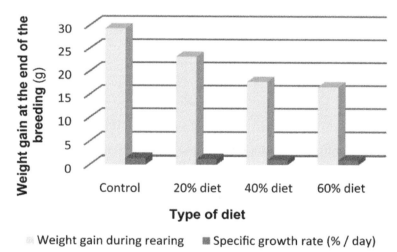

Figure 4.41. *Effect of* Gracilaria arcuata *diets on weight gain and specific growth rate of Nile tilapia (*Oreochromis niloticus*) (from Younis et al. 2019). For a color version of this figure, see www.iste.co.uk/fleurence/algae.zip*

The impact of algal diets shows an effect on the composition of muscle and carcass minerals in fish fed these diets. The observed effects are reflected by increases in the contents of these components, which are proportional to the concentrations of algae integrated into the diet.

Paradoxically, the results obtained during this study show that the addition of *G. arcuata* in the diet of a herbivorous fish like Nile tilapia has a negative effect on growth. This type of finding differs from that made with carnivorous or omnivorous fish. However, this study shows, as many do, that the best zootechnical effects are found for low levels of algae in the diet. In the case of Nile tilapia, it would therefore be interesting to study the impact on the growth of diets incorporating *G. arcuata* at levels much lower than 20%. A similar study conducted with *Ulva rigida* showed no negative effect on the growth of Nile tilapia up to a 20% level. Above this level (30%), negative effects on the apparent digestibility of the proteins were observed and thus also on the protein efficiency coefficient. The presence of fibers, which act as anti-nutritional factors and whose contents would increase according to the concentration of added algae, could be at the origin of this observation (Azaza et al. 2008).

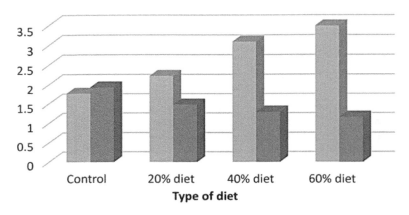

Figure 4.42. *Effect of* Gracilaria arcuate-*based diets on feed conversion rate and protein efficiency coefficient in Nile tilapia* (Oreochromis niloticus) *(from Younis et al. 2019). For a color version of this figure, see www.iste.co.uk/fleurence/algae.zip*

Paradoxically, the effect of algae in the diet of a herbivorous fish like tilapia shows a negative impact on the growth of the animal (see Table 4.16). This is in sharp contrast to the effects observed when feeding carnivorous or omnivorous fish.

Apart from macroalgae, some species of microalgae can also be involved in the nutrition of carnivorous fish. They intervene in an indirect way since they are associated with the production of rotifers (*Brachionus* sp.), which are live prey used in the nutrition of fish larvae (Muller-Feuga 2013) (Figure 4.43).

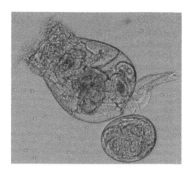

Figure 4.43. *Rotifer* Brachionus plicatilis *used in aquaculture for feeding juveniles (photo credit © Sofdrakou, personal work, CC BY-SA 4.0, https://commons. wikimedia.org/w/index.php?curid=45006975). For a color version of this figure, see www.iste.co.uk/fleurence/algae.zip*

Rotifers, used in larval rearing of fish, mainly have green microalgae belonging to the genera *Chlorella* and *Dunaliella* as a food source. Other species, including the cyanobacterium *Arthrospira platensis* (*Spirulina platensis*), are also solicited as a source of nutrients in the production of rotifers. The species involved are characterized by generally high protein and sometimes lipid contents (see Table 4.17) (Fleurence 2021a).

Microalgae use as algal feed for live prey consumed by fish larvae is not the only way this is used in aquaculture. Indeed, many species are also used for water treatment in larval rearing tanks. This type of treatment or green water treatment, following the proliferation of green algae (*Chlorella*, *Dunaliella*), favors the production of rotifers, thus inducing a beneficial effect on the growth of fish larvae (see Table 4.18). The species used are, moreover, often common in both live prey feeding and the green water treatment.

Species	Protein (%)	Carbohydrates (%)	Fat (%)
Dunaliella tertiolecta	20.0	12.2	15.0
Chlorella vulgaris	48.0	8.0	13.0
Tetraselmis suecica	31.0	12.0	10.0
Arthrospira platensis	64.0	25.0	7.0

Table 4.17. *Biochemical composition of some microalgae species and a cyanobacterium used for feeding rotifers during the rearing of fish larvae (from Becker 2013 and Fleurence 2021a)*

Species	Feeding on live prey (rotifers)	Green water treatment
Chlorella minutissima	+	+
Dunaliella virginica	+	+
Nannochloris grossii	+	+
Tetraselmis suecica	+	+

Table 4.18. *Example of microalgae species used as forage algae for live prey feeding or green water treatment (from Fleurence 2021a)*

Algae, whether macro or micro, are important nutritional inputs for fish farming. Their impact on animal growth varies according to the fish diet, the algal species involved and especially the rate of inclusion in the feed ration. The beneficial effects observed often relate to the health status of the animals (increased plasma antioxidant activity, improved immune status, resistance to pathogens). These aspects will be discussed in more detail later (see section 5.3.1).

4.6.2. Mollusks

The world production of mollusks is based mainly on the exploitation of bivalve species (oysters, mussels, clams, pectinids) and to a lesser extent on gastropod species such as abalone.

Bivalve mollusks, which are filter-feeding organisms, are produced primarily by aquaculture and their production has been increasing by 4.8% per year for the past decade (Muller-Feuga 2013). The nutrition of these organisms at the larval and spat (juvenile in the process of fixation) stage in hatcheries is based on the consumption of microalgae. The most commonly used species are *Isochrysis affinis galbana*, better known as *T-iso*, as well as some diatoms belonging to the genus *Chaetoceros* (see Table 4.19) (Fleurence 2021a).

The production of phytoplankton is therefore an essential step for the rearing of filter-feeding mollusks and more particularly of oysters (*Crassostrea gigas*, *C. plicatula*). However, this production is expensive as it can represent up to 30% of the operating cost of hatcheries.

Species	Frequency of use (%)
T-iso	72
Chaetoceros gracilis	53
Tetraselmis suecica	35
Thalassiosira pseudonana	33

Table 4.19. *Some species of microalgae used in hatcheries for filter-feeding mollusks (from Muller-Feuga 2013 and Fleurence 2021a)*

Alternative routes that are less expensive than the production of fresh microalgal biomass have been the subject of some studies (Fleurence 2021a). These new approaches are based on the use of artificial diets based on dehydrated microalgae (algal paste), micro-encapsulated products or feeds composed of yeast. These diets have very different efficiencies, and only the algal paste diet can successfully replace 50% of the fresh algal biomass in the diet of *C. virginica* oyster broodstock and spat (Fleurence 2021a).

In the same way, trials of partial or total substitution of microalgae (*T-iso*, *Skeletonema costatum*, *Chaetoceros calcitrans*) by the alga *Ulva rigida* in the nutrition of oyster broodstock (*C. gigas*) have been the subject of a recent experiment (Cardoso et al. 2019). The choice of ulva is mainly justified by the high protein and mineral contents (Fleurence et al. 2012). The properties of Ulvales to stimulate stress response and disease resistance, which are well documented in some aquaculture species (see section 5.3.1), are also the reason for the choice of this surrogate species. In this experiment, the ulva used is dehydrated and then finely ground to a size comparable to microplankton (< 150 μm).

Three alternative diets are tested (25%, 50%, 100%). The impact of these diets on the accumulation of minerals of nutritional interest and of toxic minerals in oyster flesh is part of the main objective of the study. The bioavailability of these minerals is also evaluated. *Uva rigida*-based diets have an influence on the content of trace elements. This impact varies according to the degree of substitution by the macroalga (see Table 4.20). For most of the elements studied, it is maximal, with the diet where the substitution is 100%. This influence of ulva-based diets is also observed in the accumulation of heavy metals such as lead or cadmium (see Figure 4.44). The increase in the content of these toxic minerals rises alongside the concentration of ulva used for substitution.

Element (mg/kg of MS)	0% U. rigida	25% U. rigida	50% U. rigida	100% U. rigida
Cr	–	0.91	0.92	1.73
Mn	57.20	57.70	71.10	54.40
Co	215.00	261.00	272.00	237.00
Cu	101.50	133.30	127.90	202.90
Zn	743.00	1,330.00	1,331.00	2,716.00
Se	1.14	1.50	2.12	1.95
Mo	0.45	0.35	0.23	0.63
I	1.60	2.00	2.20	2.90
Sr	18.30	26.80	26.10	35.80

Table 4.20. *Effect of* Ulva rigida-*based replacement diets on the trace element composition of oysters (from Cardoso et al. 2019)*

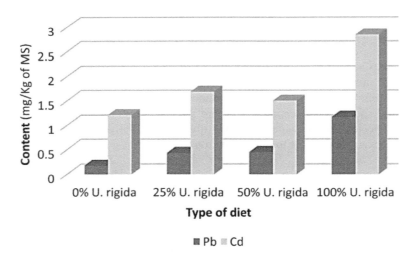

Figure 4.44. *Effect of* Ulva rigida-*based substitution diets on lead and cadmium accumulation in oysters (from Cardoso et al. 2019). For a color version of this figure, see www.iste.co.uk/fleurence/algae.zip*

The bioavailability of minerals is also impacted by Ulva-based diets (see Table 4.21). The in vitro calculated bioavailability of copper is above 100%. For strontium, notable differences are observed, but most values are close to 100%. The lowest bioavailability values are found for molybdenum and lead (56% for the 100% *U. rigida* diet). For lead, this type of finding is considered rather favorable, considering the toxicity of this element.

Element	0% U. rigida	25% U. rigida	50% U. rigida	100% U. rigida
Mn	29	23	28	16
Cu	106	112	111	110
Sr	111	91	89	73
Cd	108	97	100	73
Pb	111	63	79	56

Table 4.21. *Effect of* Ulva rigida *substitution diets on the bioavailability (%) of minerals in oyster meat (from Cardoso et al. 2019)*

The experiment carried out shows that it is possible to partially or totally substitute the microalgal fraction by a macroalgal fraction composed of *U. rigida*. This type of substitution has an impact on the accumulation of trace elements and of toxic minerals. This accumulation of minerals, as well as their bioavailability, depends on the levels of *Ulva* used in the diets.

The main source of feed for filter-feeding mollusk aquaculture remains the microalgal resource, despite attempts to replace this resource with macroalgae. The latter are solicited for the rearing of other mollusks such as abalone. As such, they contribute to the development of the halioculture.

Abalone rearing includes two important phases. The first concerns larval development, which takes place in a hatchery, and the nutrition of the larvae is based on a diet of microphytobenthos. The second phase is the development and growth of young adults, a phase that requires the addition of macroalgae.

In the larval and post-larval (juvenile) feeding phases, abalone is fed with benthic microalgae attached to the walls of Plexiglas plates placed in the development tanks. They are mainly species belonging to the genera *Navicula* and *Nitzschia,* microalgae of the diatom family (see Figure 4.45) (Fleurence 2021a). Other species such as *Cocconeis duplex* or *Tetraselmis suecica* may also be involved as fodder algae in the nutrition of abalone larvae.

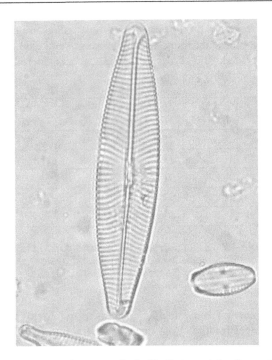

Figure 4.45. Navicula *sp. (photo credit © K. Peters, http://www.korseby.net/outer/ flora/algae/index.html, personal work, CC BY-SA 3.0, https://commons.wikimedia. org/w/index.php?curid=9432571). For a color version of this figure, see www.iste.co. uk/fleurence/algae.zip*

Adults are fed with red, green or brown macroalgae that they graze thanks to their radula. The diet of abalone may differ according to the species. Thus, the Japanese abalone (*Haliotis discus hannai*) shows an optimized growth when its diet is based on the consumption of Laminariales (brown algae). The species mainly consumed are *Laminaria* spp., *Undaria pinnatifida*, *Eisenia bicyclis* and *Ecklonia cava* (Lucien-Brun 1983). In the Japanese system, broodstock are conditioned to spawn year-round and juveniles are mass-produced in hatcheries. At this stage, benthic diatoms are used for feeding. The adults are then implanted in rocky areas, often artificial reefs, on which macroalgae have been planted to serve as a food resource for the abalone (see Figure 4.46).

In Europe, the abalone is raised for 6–12 months in nurseries. The growth of juveniles is based on a diet of diatoms, followed by the consumption of green macroalgae belonging to the genus *Ulva*. Once in the open sea, the animals are mainly fed with the red alga *Palmaria palmata*.

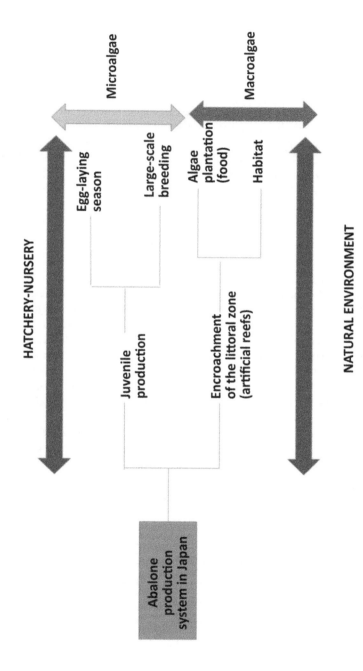

Figure 4.46. Abalone production system in Japan based on extensive aquaculture (from Lucien-Brun 1983)

Other red algae such as *Grateloupia turuturu*, a proliferating species on the French coasts, have been the subject of nutrition trials in abalone in order to replace *P. palmata*, a less available species. These trials proved to be disappointing in terms of growth on European abalone (*Haliotis tuberculata*, Garcia-Bueno et al. 2016). The abalone tested in this study showed a clear dietary preference for *P. palmata*, which is one of the species usually consumed by the animal in its natural environment. The alga *G. turuturu*, in spite of its low impact on abalone growth, is, however, of interest for its use in halioculture. Indeed, the latter appears to contain antibacterial substances that are effective against a pathogenic pest on abalone farms, the bacterium *Vibrio harveyi* (Garcia-Bueno et al. 2014) (see section 5.1.2). The development of mixed feeds containing *P. palmata* to promote animal growth and *G. turuturu* to improve its resistance to the *Vibrio* pathogen appears, with regard to this study, to be a pertinent strategy for the development of European abalone halioculture.

4.6.3. Crustaceans

Crustacean farming, particularly shrimp farming, is expanding rapidly. According to the FAO and OECD, aquaculture of this organism is expected to progress by 35% within a decade (Fleurence 2021a). Aquaculture of shrimp at the larval stage relies on the use of forage microalgae. The species mainly used are diatoms of the genus *Chaetoceros* or the green alga *Tetraselmis suecica* in the case of the Japanese shrimp *Penaeus japonicus*. These species significantly improve larval growth and survival rates (Muller-Feuga 2013). However, this finding is not applicable for all shrimp species. Indeed, depending on the algae used, notable differences can appear in terms of larval survival. This is the case for the giant tiger shrimp (*Penaeus monodon*) whose larval survival rate is about 24% in the presence of *Dunaliella tertiolecta* compared to 66% in the presence of *Tetraselmis chuii* (see Figure 4.47). Algae of the genus *Chaetoceros* or *Tetraselmis* are also used for the nutrition of crab larvae, another crustacean subjected to aquaculture production (see Table 4.22).

Species	Shrimp larvae	Crab larvae
Chaetoceros calcitrans	++++	++
C. muelleri/C. gracilis	++++	
Skeletonema sp.	++++	
Tetraselmis sp.	++	++
Nannochloropsis sp.		++++
Phaeodactylum sp.		++++

Table 4.22. *Examples of microalgae used as a food source in the nutrition of crustacean (shrimp, crab) larvae (from Zmora et al. 2013)*

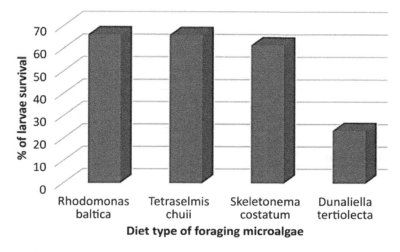

Figure 4.47. *Survival rate of giant shrimp (*Penaeus monodon*) larvae in function of microalgae species used in larval nutrition (from Kurmaly et al. 1989 and Fleurence 2021a)*

Microalgae are therefore essential nutritional inputs for the aquaculture production of shrimp larvae. Macroalgae can also be used in the feeding of these animals, but at a later stage (juveniles, adults). In particular, they are used in shrimp co-culture to limit the use of artificial feed (Cruz-Suarez et al. 2010). In this type of aquaculture model, the algae are attached to the submerged mesh in the shrimp rearing tanks and the contribution of the artificial feed is progressively decreased (100%, 90%, 55%) until it is absent from the supplied feed ration.

An experiment using the green alga *Ulva clathrata* and the shrimp species *Litopenaeus vannamei* (Pacific shrimp) showed interesting results in both growth and quality of the crustacean's body. The consumption of algae by shrimp increased the conversion efficiency of the artificial feed provided during rearing as well as the growth rate. This resulted in a decrease in artificial feed intake by 10–45% and an improvement in growth rate by nearly 60%. Ulva feed intake in the presence or absence of artificial feed also had an impact on the lipid composition of the flesh (see Figure 4.48) as well as on its organoleptic quality (see Table 4.23). In the absence of artificial feed, the ulva feed intake became optimal and the lipid content of the crustacean flesh was minimal (see Figure 4.48). Co-culture with ulva in the absence of artificial feed had a positive effect on the eicosapentaenoic (C25:5n-3) and docosahexaenoic (C22:6n-3) acid content (see Figure 4.49).

Furthermore, co-culture improves the pigmentation of shrimp, suggesting that the carotenoids present in ulva, and more particularly lutein, are easily assimilated and metabolized by the animals.

Diet	Redness	Yellowness	Tenderness
0% Artificial feed	16.8	34.0	65.2
55% Artificial feed	16.2	30.7	69.1
90% Artificial feed	11.5	26.3	73.7
100% Artificial feed	12.1	29.1	71.6

Table 4.23. *Effect of nutritional feed on the organoleptic quality of* Litopenaeus vannamei *shrimp meat in a "shrimp-Ulva" co-culture system (from Cruz-Suarez et al. 2010)*

Other studies concerning the feasibility of algal co-culture coupled with shrimp farming are available in the literature. They are based on the use of green algae such as *Caulerpa sertularioides* (Porchas-Cornejo et al. 1999) or red algae like *Kappaphycus alvarezii* (Lombardi et al. 2006). Some show that algal co-culture increases by three times the growth rate of shrimp. This is particularly the case for co-culture with *C. sertularioides*. On the other hand, other trials show the absence of negative interference between algal co-culture and shrimp rearing; this translates into a neutral influence on the growth of the animals, their survival rate and the conversion rate of the feed provided. The latter case is observed when co-cultured with *K. alvarezii*.

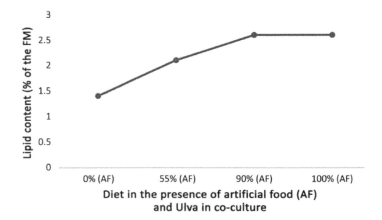

Figure 4.48. *Effect of the rate of artificial feed in the diet of shrimp grown in co-culture with ulva on the lipid content of the meat (from Cruz-Suarez et al. 2010)*

Apart from the integration of macroalgae in the indirect feeding of shrimp via the co-culture process, some works have shown the feasibility of using fresh macroalgae for the nutrition of *L. vannamei* larvae (Cruz-Suarez et al. 2010). These algae introduced in a formulated feed have a very positive effect on growth (size, weight) and survival rate of the animals (48–53%), which are also compared to other diets (dehydrated algae and artificial feed, dehydrated algae alone).

The coupling of seaweed culture and shrimp farming is a production method that allows the introduction of macroalgae in the diet of these crustaceans. The effects obtained differ according to the species. The addition of algae in the rearing tanks presents several interests. Indeed, macroalgae are efficient purifiers of effluents generated by aquaculture. This is particularly true for ulva which are natural traps for inorganic nitrogen compounds such as nitrates, nitrites or ammonium. Algal co-culture thus contributes to maintain a water quality which is compatible with the health of the livestock. This approach developed for the culture of other aquaculture organisms (mollusks, crustaceans, echinoderms) is at the origin of the concept of integrated aquaculture.

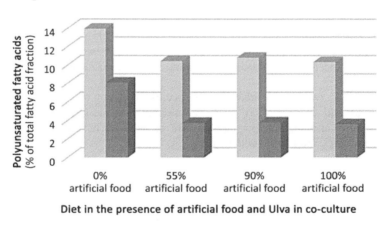

Figure 4.49. *Impact of ulva co-culture and artificial feed provision on eicosapentaenoic (EPA) and docosahexaenoic (DHA) acid content in shrimp meat (from Cruz-Suarez et al. 2010). For a color version of this figure, see www.iste.co.uk/fleurence/algae.zip*

Co-culture also has the advantage of limiting the supply of artificial feed and thus reduces the economic and environmental cost of conventional aquaculture.

The mariculture of crabs, or carcinoculture, represents a quasi-industrial activity in Asia. With a production of 800,000 tons, China is the world's leading producer of this crustacean. The main species produced is *Eriocheir sinensis,* marketed under the commercial name of Chinese crab (Fleurence 2021a). The larvae (zoe) and their post-larval stages (megalops) are reared in tanks and fed with a complex diet composed of zooplankton, Artemia, vitellus and microalgae. The algal species used belong preferably to the genera *Chaetoceros* or *Chlorella*.

The Chinese crab is then grown in a natural environment physically limited by natural barriers (lakes, ponds, rice fields) or artificial barriers (nets). The animals are fed by taking resources available from the natural environment as well as those incidentally provided by the aquaculturist (rapeseed meal, barley, corn, slaughterhouse waste, silkworm chrysalis, etc.).

In this context, the supply of macroalgae in a direct manner or in co-culture is little requested.

4.6.4. *Echinoderms*

Sea urchin farming or echinoculture is a response to the scarcity of this resource in the natural environment, particularly in Japan, a major consumer of this animal. In Europe, the stocks of sea urchins (*Paracentrotus lividus*) present on the French and Irish coasts collapsed in the 1980s and the resource has never recovered.

The *Strongylocentrotus droebachiensis* species, found on the east coasts of the United States and Norway, has been commercially successful in the Japanese market, but its variable and low egg content as well as its dull color limit the development of any interesting added value.

In order to overcome these disadvantages, the rearing of different species of sea urchins has been developed in extensive and non-intensive ways. The rearing process includes a larval production phase (pluteus larva) followed by the growth phase of juveniles into adults. Larval feeding is based on the supply of microalgae, some of which belong to the genera *Dunaliella, Chaetoceros* or even *Isochrysis* (Brundu et al. 2016). The metamorphosis of the larva that marks the beginning of the post-larval stage is a crucial and sensitive step in the rearing process. It is provoked by the presence or supply of inducers that can be microalgae, biofilms, macroalgae or even conspecific adults (see Table 4.24). The stimulation is then done with or without contact.

The percentage of metamorphosis evolves over time to reach a maximum about three to four days after induction (see Figure 4.50). It is significantly higher in the presence of the natural biofilm compared to that recorded with the *Ulvella lens* biofilm.

Inductor	Nature of the induction	% metamorphosis	% survival
Cocconeis scutellum (Diatom)	With contact	63	60
Natural biofilm (Benthic diatoms)	With contact	65	90
Ulvella lens	With contact	30–35	95
Ulva rigida	With contact	38	34
Laminaria digitata	With and without contact	3–20	–
Corallina elongata	No contact	80	55
Adult sea urchins	No contact	20	–

Table 4.24. *Effect of different inducers (biofilms, microalgae, macroalgae and adult sea urchins) on the initiation of metamorphosis in sea urchin (*Paracentrotus lividus*) larvae (from Castilla Gavilan 2018)*

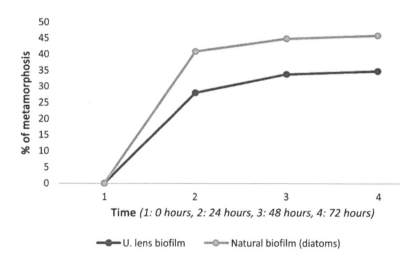

Figure 4.50. *Effect over time of a natural or* Ulvella lens *biofilm on the percentage of metamorphosis of sea urchin larvae (*Paracentrotus lividus*) (from Brundu et al. 2016)*

In addition to the percentage of metamorphosis, the post-larval survival rate is also a critical parameter for rearing success. In the *P. lividus* species, survival to metamorphosis can vary depending on the nature of the metamorphosis inducers or microalgae species used for larval feeding (see Table 4.24 and Figure 4.51). When larvae are fed a *Dunaliella–Chaetoceros–Isochrysis* mixture, this survival rate is particularly low (4.2%). The influence of diets on post-larval survival remains very strong for this species. Indeed, after 180 days post-metamorphosis and post-benthic settlement, only juveniles that have received feeding at the larval level based on *Dunaliella tertiolecta* or the *Chaetoceros–Dunaliella* mixture survive (Brundu et al. 2016).

A few days after metamorphosis, the juveniles are fed with fresh macroalgae. This mode of nutrition can be completed by the contribution of an artificial food if the resource of macroalgae is limited during the adults' growth phase.

Otherwise, the adults are mainly fed through the preferential supply of red algae such as *Palmaria palmata* or *Chondrus crispus*.

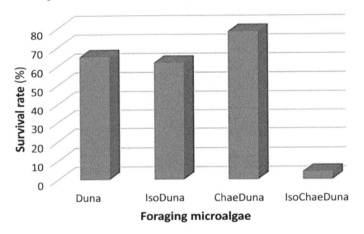

Figure 4.51. *Effect of microalgae diet on larval survival of* Paracentrotus lividus *species during larval metamorphosis (Duna:* Dunaliella*; IsoDuna:* Isochrysis–Dunaliella *mixture; ChaeDuna:* Chaetoceros–Dunaliella *mixture; IsoChaeDuna:* Isochrysis–Chaetoceros–Dunaliella *mixture) (from Brundu et al. 2016)*

The culture of sea urchins is a difficult activity, limited by an essential stage, which is the passage from the larval state to the juvenile one (metamorphosis). The use of microalgae and macroalgae is essential for the development of echinoculture. They intervene in the process of development and growth as both food and inducers of the metamorphosis process.

Holothurians, echinoderms which are known as sea cucumber or sea spade, are traditionally consumed in China. They are considered to be "health food" in Chinese popular medicine. Of the 20 or so species on offer for consumption, only one is farmed. This is the *Apostichopus japonicus* (Chen 2003) species. In the hatcheries, the larvae of *A. japonicus* are fed with microalgae such as *Dunaliella salina*, *Phaeodactylum tricornutum* and *Chaetoceros simplex* or with marine yeast.

As in the case of sea urchins, diet influences the survival rate of larvae and juveniles (see Figure 4.52).

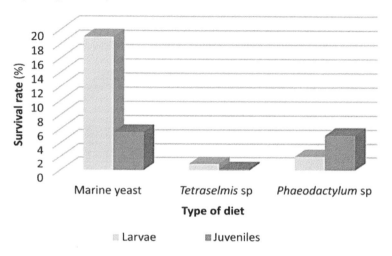

Figure 4.52. *Effect of different diets based on microalgae or marine yeast on the survival rate of larvae and juveniles of the sea cucumber* Apostichopus japonicus *(from Chen 2003). For a color version of this figure, see www.iste.co.uk/fleurence/algae.zip*

When the juveniles reach 2–3 cm, they are transplanted into seawater basins where they can complete their growth. These basins are previously coated so that the benthic macroalgae can be used as food for the holothurians.

Algae, whether micro or macro, are already used in animal nutrition. In this respect, their use is far from being innovative. However, the frequency of their use is low compared to other food sources. The use of plant proteins derived from the intensive cultivation of species such as soybeans or corn, which require a lot of fresh water, currently poses a problem in terms of the eco-responsibility of animal production chains. The use of soy proteins in the formulation of feeds for fish farming also raises questions about the relevance of such a choice (Fleurence and Guéant 1999). This is especially true since many species of red algae (*Palmaria palmata*, *Porphyra yezoensis*) or green algae (*Ulva lactuca*) have equivalent or even higher protein contents (>25% of dry matter) than many legumes (Fleurence 2004).

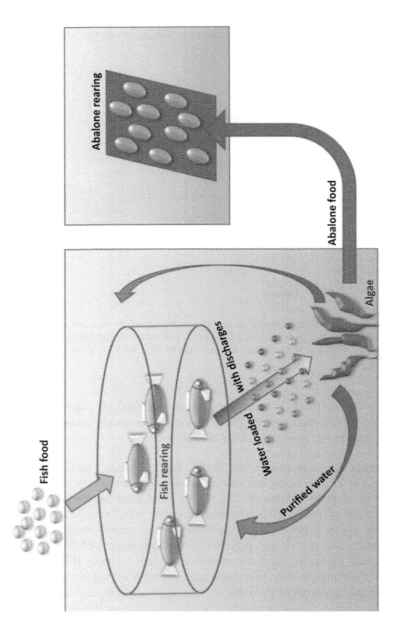

Figure 4.53. *Example of an integrated aquaculture system in which algae are used as a purifier of fish effluents and as a food resource for abalone farming (source: Y.F. Pouchus, 2022)*

Algae can also be used in aquaculture to purify fish farms and as a food resource for other species of aquacultural interest (abalone, echinoderms). This concept of an integrated aquaculture system is of both ecological and economic interest (see Figure 4.53).

In addition, algae contain substances with antiparasitic and immune-stimulating activities that make them a valuable aid to animal health, thus limiting the use of antibiotics in productive livestock systems (see section 5.1.2).

5

The Biological Activities of Algae in Plant or Animal Health

Algal extracts or molecules are often described as having biological activities that are of interest for plant or animal health. These activities are mainly antiparasitic, antibiotic, antiviral or antifungal. In addition, algal compounds are likely to induce the natural defense mechanisms of organisms. For plants, this translates into the activation of the hypersensitivity mechanism (see section 5.2.1). In animals, these compounds can activate the immune system, either acquired or innate (complement system) (see section 5.3).

5.1. Antiparasitic and antimicrobial activities

5.1.1. *Plant parasites and pathogens*

A parasite is an animal or plant organism that colonizes another organism, called a host, and lives at its expense. The parasite can be a unicellular organism (protozoan, protophyte) or multicellular. Most of the time, it generates damages in the tissues or the root system of the plants, leading to a weakening of the latter.

The term "pathogens" refers to microorganisms (viruses, bacteria, fungi) whose penetration into the organism will generate a severe pathology.

Pests attacking plants, and more particularly crops of food interest, are numerous and varied in nature. Among the main ones, we can mention aphids, cochineals, mites or nematodes. Nematodes are small worms, present in the soil, which attack the root system of plants (see Figure 5.1). These parasites induce the formation of root galls that will disturb the absorption of water and nutrients in the soil.

The crops most susceptible to this pest are tomatoes, eggplant, potatoes and melons. Damage related to this type of parasitism can bring on 10–25% of losses in crops (Veronico and Melillo 2021).

The effects of the application of seaweed extracts on the nematode infestation of tomato plants are well documented. A study on the use of a concentrated extract of *Ecklonia maxima* (Kelpack) so as to limit nematode *Meloidogyne incognita* reports interesting results (see Figure 5.1) (Crouch and van Staden 1993). In this study, the extract was applied by drenching to saturate the soil where tomato (*Lycopersicum esculentum*) seedlings were located or also by foliar spray. The foliar application method was not very effective for the growth of nematode-infested plants and had little effect on reducing the number of root galls. In contrast, the application of algal extract (1% concentration) by soil drench improved plant growth in both nematode-free and nematode-infested soils. It also limited the number of galls generated at this concentration (–40%). The same experimentation was carried out on two tomato cultivars, one being sensitive (Rana) and the other resistant (M1) to the nematode. The susceptible cultivar facilitated nematode reproduction, while the resistant one limited it.

Figure 5.1. *Young nematode (*Meloidogyne incognita*) entering the root of a tomato plant (photo credit © W. William Wergin and R. Sayre, colored by Ausmus S., United States Department of Agriculture, d2549-1, CC BY 2.0, https://commons.wikimedia. org/w/index.php?curid=26453455). For a color version of this figure, see www.iste.co. uk/fleurence/algae.zip*

On the Rana cultivar, the number of root galls was significantly reduced in the presence of the extract, regardless of the concentration used (0.2%, 0.4%) (see Figure 5.2). In contrast, for the resistant cultivar, the application of the extract at a concentration of 0.4% significantly increased the number of root galls compared to the control and plants treated with a concentration of 0.2%. This result is confirmed by the egg count, which shows a higher census of these structures in resistant plants treated with the algal extract with a 0.4% concentration.

The response of the resistant cultivar to the application of the concentrated algal extract (0.4%) could be explained by the action of exogenous and endogenous growth regulators (phytohormones) on the plant. Indeed, the exogenous application of kinetin and auxin on resistant *Meloidogyne* tomato cultivars is known to favor root production of giant cells and the formation of galls. The 0.4% extract, because of its phytohormone concentration, probably altered the nematode resistance mechanism. In contrast, this mechanism seems to be less affected with an algal extract with half the concentration.

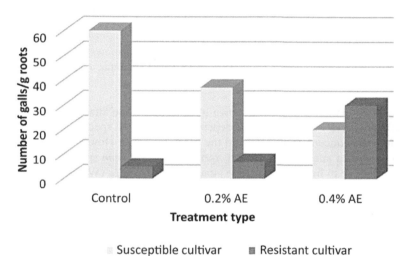

Figure 5.2. *Effect of brown algal extract* Ecklonia maxima *(AE) on the number of root galls generated by the nematode* Meloidogyne incognita *on susceptible and resistant cultivars of tomato (*Lycopersicum esculentum*) (from Crouch and van Staden 1993)*

Soil treatment with ethanolic extracts (1 mg/mL) of green (*Ulva fasciata*) or red (*Corallina officinalis, C. mediterranea*) algae was also effective in limiting *M. incognita* infestation of tomato plants (Veronico and Melillo 2021). The *U. fasciata* extract appeared to be the most active, reducing the number of galls per root by almost 78% compared to 50% and 35% with the *C. officinalis* and *C. mediterranea* extracts, respectively.

The mechanism of action of algal extracts on the nematode differed among the algal species tested. This mechanism can be the inhibition of egg hatching or the induction of juvenile mortality during the reproductive and developmental cycle of the parasite (see Table 5.1). In this case, the algal extract disrupted the completion of the cycle and controlled the reproduction of the parasite, thereby limiting the damage to the host.

In some species and more particularly in brown algae, a nematicidal activity is also possible as a mechanism of action. This is the case of the aqueous or methanolic extracts of *Stoechospermum polypodioides*, *Sargassum tenerrimum* or *swartzii*, which show a strong nematicidal potential against the nematode *M. javanica* (Veronico and Melillo 2021). On the model plant *Arabidopsis thaliana*, treatment with an alkaline extract of *Ascophyllum nodosum* was also found to be effective in reducing the number of female nematodes belonging to the species *Meloidogyne javanica* (–40%) and in limiting their fecundity (–42% of eggs produced) (Wu et al. 1998).

Apart from tomato plants, seaweed extracts were also tested on many crops targeted by nematodes. These were fruit crops (okra, banana, lemon) or vegetable crops (eggplant) (see Table 5.1).

Many algal extracts have been shown to limit nematode attacks (see Table 5.1). Their efficacy and mode of action depend on the algal species, the method of application, the concentration used and the crops involved. However, a recent study based on metadata analysis shows that *Ascophyllum nodosum* species appears to be the most effective in limiting nematode attacks on most crops (Williams et al. 2021).

The antihelminthic activities of algae on nematodes are now shown. The mechanisms of action on the reproductive cycle and on the biocidal activity towards the worm are partially known. Further studies on the subject should allow the development of an active product of natural origin and specific to the parasite. This approach could then limit the use of chemical pesticides that are harmful to animals or humans.

Thrips, which are small insect pests, as well as phytophagous mites, cause considerable damage to vegetable, fruit and flower crops worldwide. These plant tissue-consuming parasites are also vectors of viral diseases. A commercial extract of *Ascophyllum nodosum* has been shown to be effective in the field in suppressing the deleterious effects of thrips and mites (Holden and Ross 2012). The algal extract employed reduces the number of thrips (*Scirtothrips perseae*) colonizing avocado trees by 68%. This efficacy is comparable to that of abamectin, a synthetic insecticide used against this pest. The extract is also very active against the avocado mite (*Oligonychus perseae*), since it leads to an 87% reduction of the parasites present on the leaves of the avocado tree. As before, this efficacy is similar to that reported for abamectin.

Algal species	Nematode	Concentration of the extract	Targets treated	Plant hosts
Ascophyllum nodosum (brown algae)	*M. javanica*	3.6%	Juveniles	Tomato
Ascophyllum nodosum (brown algae)	*M. javanica*	1%	Eggs	Basil
Ascophyllum nodosum (brown algae)	*Radopholus similis*	2.24 kg/ha (powder)	Natural soil	Lemon tree
Cystoseira myrica (brown algae)	*M. incognita*	20 g/plant (powder)	Juveniles	Eggplant
Sargassum tenerrimum (brown algae)	*M. javanica*	0.5–1% (weight/volume) (powder)	Eggs	Combo
Sargassum vulgare (brown algae)	*Meloidogyne* spp.	5 g/kg of soil	Juveniles	Banana tree
Ulva lactuca (green algae)	*Meloidogyne* spp.	5 g/kg of soil	Juveniles	Banana tree
Halimeda tuna (green algae)	*M. javanica*	1% (weight/volume)	– Eggs – Juveniles	– Sunflower – Tomato
Digenea simplex (red algae)	*M. incognita*	20 g/plant (powder)	Juveniles	Eggplant

Table 5.1. *Some examples of the use of extracts of brown, green or red algae for their antinematode activities on the crops of agronomic interest (according to Veronico and Melillo 2021)*

The same result was obtained on the red spider (*Tetranychus urticae*), which is phytophagous on strawberry plants (*Fragaria* sp. variety Redgaunlet). This small pest, also known as a "two-spotted spider mite", is one of the most common pests of many greenhouse or tunnel crops.

Treatment with Maxicrop, a commercial product with *Ascophyllum nodosum* as its basis, suppresses the build-up of populations of this pest in strawberry crops grown in tunnels (Hankins and Hockey 1990).

Aphids are hemipteran insects that attack many crops, especially Brassicaceae. They cause damage to plants by stings and by the production of honeydew leading to the degeneration of leaves. The best known representative of this family of pests is the cabbage aphid (*Brevicoryne brassicae*). A phenolic compound found in the brown alga *Ecklonia maxima* has been shown to be effective in limiting the damage caused by aphid infestation of cabbage plants (*Brassica oleracea* variety Drumhead) (Rengasamy et al. 2016). This compound belongs to the eckol family (see Figure 5.3) and is extracted from the commercial product Kelpak.

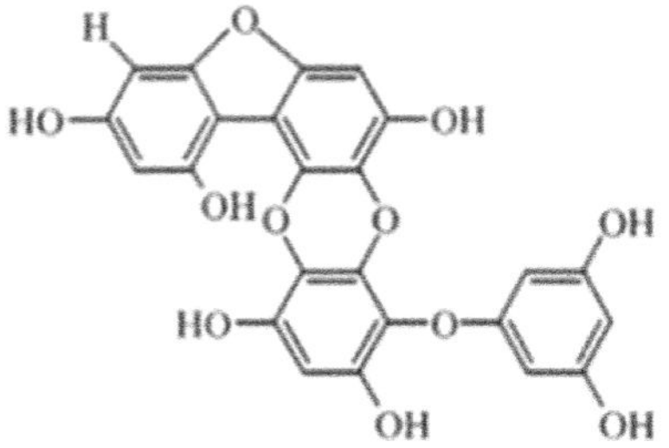

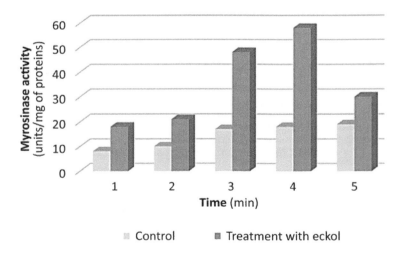

Figure 5.3. *Structure of eckols contained in brown algae of the genus* Ecklonia *(Fleurence and Ar Gall 2016)*

Figure 5.4. *Impact of foliar spray of eckol solution (10^{-6} M) on the myrosinase activity of cabbage plants (from Rengasamy et al. 2016)*

Eckol extracted from *E. maxima* is applied by foliar spray. A significant improvement of the growth parameters and photosynthetic activity of the plants is reported during this treatment. This polyphenol also has a stimulating effect on myrosinase activity (see Figure 5.4). This enzyme, mainly found in the Brassicaceae family, plays an important role in the defense mechanisms of plants. In particular, it hydrolyzes glucosinolates into isothiocyanates, which are chemical repellents for phytophagous insects because of their pungent taste. Eckol is also effective as a biocide. At a concentration of 10^{-5} M, this molecule induces a mortality rate in the aphid population that is close to 90%, which is remarkable for a product of natural origin (see Figure 5.5).

This work shows that a molecule called eckol, obtained from a commercial extract of *Ecklonia maxima*, has two duties: it acts as a biostimulant and as a natural insecticide against a phytophagous parasite. This double functionality of an algal compound is not very frequent. It opens up real prospects for the use of algae as a substitute for chemical fertilizers and pesticides.

In addition to animal pests, crop plants can be confronted with plant pests. Among these is the rambling broomrape (*Phelipanche ramosa* or *Orobanche ramosa*) (see Figure 5.6), a non-chlorophyllous, holoparasitic plant, which causes extensive root damage to crops of economic interest such as tobacco (*Nicotiana tabacum*), rapeseed (*Brassica napus*), sunflower (*Helianthus annuus*) or tomato in tropical regions (*Lycopersicum esculentum*).

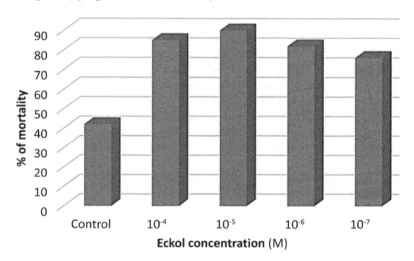

Figure 5.5. *Impact of different concentrations of eckol on cabbage aphid mortality (from Rengasamy et al. 2016)*

Figure 5.6. *Parasitic plant of broomrape (*Phelipanche ramosa*) (photo credit © J.-B. Pouvreau 2021). For a color version of this figure, see www.iste.co.uk/fleurence/algae.zip*

The means of control against this parasite show a limited effectiveness (chemical pesticides, genetic control, pulling up of plants before the flowers appear). Genetic control is based on the selection of varieties resistant to broomrape. This approach has led to the selection of sunflower varieties with genes for resistance to the parasite, known as OR1, OR2, OR3, OR4, etc.). Unfortunately, this genetic approach is often bypassed by the appearance of increasingly aggressive populations of broomrape escaping this alternative pathway. In this context, the use of seaweed extract as a means of control was the subject of a study published in 2007 (Economou et al. 2007). The latter is based on the use of a commercial product Algit Super developed from an *Ascophyllum nodosum* extract. The product was tested at different concentrations (2.5–1.2 10^{-3} v/v), and its impact on the germination of *Orobanche ramosa* seeds was evaluated. The germination process was dependent on the concentrations used. It was stimulated for low concentrations (C6: 0.038 v/v) and inhibited for high concentrations (C1: 2 v/v) (see Figure 5.7).

Stimulating the germination of broomrape under certain conditions can be an effective means of controlling this parasite. Indeed, an early emergence of broomrape before crop establishment would be equivalent to a "suicide germination" for this non-chlorophyllous plant that can only live at the expense of its host.

In contrast to animal parasites, the number of studies involving algal extracts in the control of parasitic plants appears to be very limited. An extension of this type of approach involving other algal extracts becomes an interesting research issue today.

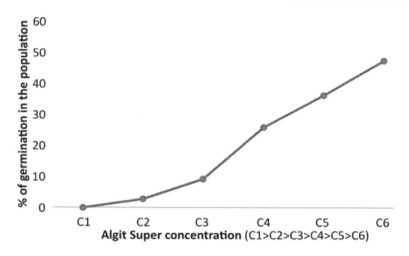

Figure 5.7. *Effect of the product Algit Super (*Ascophyllum nodosum*) depending on the concentrations used on the germination of* Orobanche ramosa *seeds (C1: 2 v/v; C2: 1.25 v/v; C3: 0.3 v/v; C4: 0.15 v /v; C5: 0.075 v/v; C6: 0.038 v/v) (from Economou et al. 2007)*

Cultivated plants also have to deal with numerous pathogens that cause diseases and yield losses in crops. These pathogens are bacteria, microscopic fungi (micromycetes) or viruses. Faced with an attack by these microorganisms, plants can respond by triggering natural defense mechanisms. The hypersensitivity reaction is one of these inducible mechanisms. Considered as a plant immunity response, this reaction is treated separately in this book (see section 5.2). Algal extracts or molecules often appear as activators or elicitors of the hypersensitivity response. Their activities against the pathogen can be considered, in this case, as an indirect action.

In addition, some seaweed extracts have been shown to have a direct action on the pathogen by presenting an antibiotic, antifungal and even viricidal activity.

Bacteria of the genus *Xanthomonas* are plant pathogens that attack many crops (bean, alfalfa, rice, cotton, soybean, pepper, cherry, peach, lemon). The bacterium *Xanthomonas axonopodis* is responsible for citrus canker, a devastating disease causing weakening of trees, as well as premature leaf and fruit fall. Some algal extracts obtained from species collected from Indian coasts are found to possess antibiotic activity against *X. axonopodis* pv. *citri* (Hasse) species (Arunkumar and Sivakumar 2012). The algal fractions which were tested were obtained by extraction in a chloroform–methanol (1:1) mixture. Extracts are derived from algae collected seasonally (post-monsoon, summer, pre-monsoon, monsoon) to assess the presence of a possible temporal variation in antibiotic activities. The highest antibacterial activity is reported for extracts obtained from the red alga *Portieria hornemannii*. This activity varies seasonally, and is optimal in algae harvested during the monsoon season (November) (27% inhibition of bacterial growth) and is associated with five different compounds (see Figure 5.8).

Regardless of the algae tested, the antibacterial activity of the extracts appears to be lower during the summer and pre-monsoon periods (August) (Arunkamar and Sivakumar 2012). This variation is probably one of the consequences of the secondary metabolism of the algae which fluctuates with the seasons. Through this, it gives an indication of the chemical nature of the bioactive compounds which are probably secondary metabolites of polar nature because they are extractable in methanol.

This study highlights the presence in algae of compounds that act directly on a pathogenic bacterium responsible for numerous ravages on crops of agronomic interest. In this, it proposes an alternative to the use of chemical pesticides.

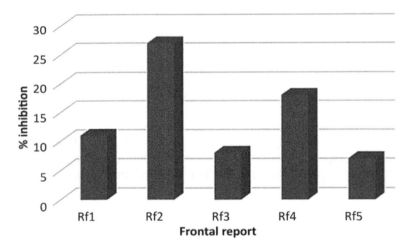

Figure 5.8. *Antibacterial activities against* Xanthomonas axonopodis *of the substances present in an extract of the red alga* Portieria hornemannii *harvested during monsoon. The active compounds are characterized by their migration in chromatography on thin layer (Rf1: 0.19; Rf2: 0.23; Rf3: 0.37; Rf4: 0.57; Rf5: 0.62) (from Arunkumar and Sivakumar 2012)*

Fungal diseases due to phytopathogenic fungi are numerous and varied. They are mainly associated with basidiomycetes (genus *Puccinia*) or ascomycetes (*Fusarium* or *Botrytis* genera). In wheat, the *Puccinia graminis* species is responsible for the black rust disease. Species of the genus *Fusarium* are the cause of many cryptogamic diseases affecting vegetable and ornamental plants. The main symptoms of Fusarium diseases are stem wilting, yellow spots and root rot. They are often referred to as "foot disease" by the general public.

Fungi of the genus *Botrytis* also generate numerous cryptogamic diseases. The species *Botrytis cinerea* is responsible for the disease known as "grey rot". This disease affects many crops of agronomic interest such as grape vine, sunflower, tomato or strawberry. Numerous studies report the impact of seaweed extracts and more particularly of *Ascophyllum nodosum* on the control of these diseases. The effects observed are mainly related to the activation of biochemical defense mechanisms in treated plants and thus to an indirect activity of the algal extract on the pathogen. As before, these aspects are further discussed in section 5.2.

Independently of this, some works have been able to associate a direct antifungal activity with the extracts of green algae (*Enteromorpha flexuosa, Chaetomorpha antennina, Ulva lactuca*) (Chanthini et al. 2012). Methanolic and ethyl acetate extracts of these algae have indeed been shown to have antifungal activity in vitro on spores of the fungus *Alternaria solani* (see Figure 5.9). This phytopathogen is responsible for Alternaria leaf blight and can affect tomato plantations in tropical areas, as well as potato plantations.

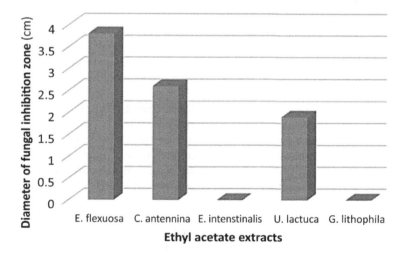

Figure 5.9. *In vitro antifungal activity on* Alternaria solani *spores of different ethyl acetate extracts obtained from different macroalgal species (*Enteromorpha flexuosa, Chaetomorpha antennina, Enteromorpha intestinalis, Ulva lactuca, *Grateloupia lithophila) (Chanthini et al. 2012)*

Algal extracts, and more particularly those of brown algae (*Ecklonia maxima*, *Ascophyllum nodosum*), are often described for their pest control, or even biocidal, activities against certain phytopathogens such as fungi or bacteria. This direct action is often associated with methanolic or alkaline extracts. However, the bioactive compounds of these extracts are rarely characterized from a chemical or structural point of view. This is not the case for eckols, which are well documented for their antiparasitic activities.

The action of algal extracts and in particular that of polysaccharides such as laminarin, carrageenans or fucoidans in the protection of plants against parasites and pathogens, including viruses, is, however, well documented. These extracts or algal molecules elicit the plants' biochemical resistance mechanisms and participate in the reinforcement of their immune status (see section 5.2).

5.1.2. Animal parasites and pathogens

Epizootics are epidemics that affect terrestrial and aquatic animal production. When these epidemic phenomena affect a continent or the world, the accepted term is panzootic. If they strike a region in a constant manner, we would use enzootic. When the infection is transmissible to humans (tuberculosis, plague, avian flu, rabies), the term anthropo-epizootic is used.

Animal epidemics are associated with infection by fungal, bacterial or viral pathogens. Some plathelminth worms (tapeworm, small sheep liver fluke) sometimes present in pig or sheep farms are formidable parasites. They can also contaminate humans through certain eating habits (undercooked meat or vegetables).

Numerous external parasites impact cattle breeding (flies, lice, scabies, ticks). Internal parasites are mainly nematode worms (strongyles) or plathelminths (flukes). The interest of using algae or their extracts in the research of a vermifuge or insecticide activity against these parasites has been little evaluated in ruminants and pigs. This is in contrast to crop plants that are subject to helminthic or insect attack (see section 5.1.1). In contrast, algal extracts are often tested for their antibiotic activities and impact on animal health.

The effect of algal extracts on certain pathogens or parasites of aquaculture species is well known and is the subject of numerous studies. Most of the time, these studies concern the antibacterial or even antiviral activity of algal extracts. In this case, the direct effects of algal extracts on pathogens are studied. Besides this aspect, the studies carried out can also focus on the contribution of the algae in the activation of resistance mechanisms of the animals to infections (see section 5.3.1).

In this case, it is mainly a question of evaluating the induced effects of algae on animal health.

The main bacterial pathogens involved in diseases affecting species produced by aquaculture belong to the genera *Vibrio*, *Aeromonas* and *Photobacterium* (formerly *Pasteurella*) (see Table 5.2).

Bacterial germ	Targeted species	Disease	Symptoms
Vibrio harveyi	Abalone	Vibriosis	Tissue necrosis
Vibrio spp.	Oysters, clams, scallops, farmed shrimp	Vibriosis	Black lesions in soft tissue or cuticle for shrimp
Aeromonas salmonicida	Salmonidae	Furunculosis	– Redness in the fins – Blood-laden boils
Photobacterium piscicida	Bass, bream, turbot, mullet, porgy	Pasteurellosis	– Hemorrhagic outbreaks – Blackening of the skin

Table 5.2. *Examples of major bacterial diseases affecting aquatic species that can be produced by aquaculture*

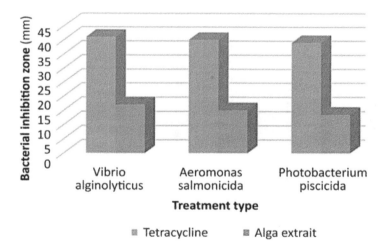

Figure 5.10. *Comparison of in vitro antibacterial activities between a synthetic antibiotic (Tetracycline (30 µg)) and an ethanolic extract of* Asparagopsis taxiformis *(2 mg) against aquaculture pathogens (from Genovese et al. 2012). For a color version of this figure, see www.iste.co.uk/fleurence/algae.zip*

Ethanolic extracts of the red alga *Asparagopsis taxiformis* are described in the literature as having antibacterial activity against many marine pathogens (Genovese et al. 2012). This in vitro activity, however, and is lower than those reported for conventional antibiotics (see Figure 5.10). It is highest against *Vibrio alginolyticus*, which is a common pathogen of aquatic animals and of the human species (17 mm zone of bacterial inhibition).

Apart from *A. taxiformis*, extracts of other red algae have also been shown to have antibacterial activities, particularly towards *V. harveyi*. This is the case, for example, for the species *Grateloupia turuturu*, whose aqueous extract shows a maximum inhibition rate of *V. harveyi* of 16% (Garcia-Bueno et al. 2014). This result corroborates that described for *A. taxiformis*, which establishes a maximum in vitro anti-vibrio activity of 17% (Genovese et al. 2012). In contrast, the antibacterial activity of *G. turuturu* varies strongly with the seasons (see Figure 5.11). It is maximal in the spring, almost nil in summer and low in winter. This notion of seasonal variability associated with the antibacterial activities of algal extracts is important to take the studies on the subject further. Indeed, it is the key that will allow a valorization of algae as a source of "natural" antibiotics and will open an alternative to the use of chemical products for veterinary use.

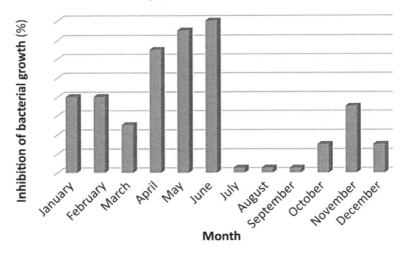

Figure 5.11. *Seasonal variation of antibacterial activity (*Vibrio harveyi*) of an aqueous extract of the red alga* Grateloupia turuturu *(from Garcia-Bueno et al. 2014)*

In addition to antibiotic activities, some studies report the presence in algal extracts of antiviral activity against certain viruses that are dangerous for salmonids (infectious hematopoietic necrosis virus (IHNV), infectious pancreatic necrosis virus (IPNV)) (Kang et al. 2008). Methanolic extracts obtained from green, brown or red algae have been screened for viricidal activity and protective effect against splenic cells. The viricidal activity is calculated from the evaluation of the TCID 50 (Tissue Culture Infective Dose). The measurement of this parameter gives the viral titer in the infected tissues. This titer is determined when 50% of the cultured cells show cytopathological signs.

The viral dose inoculated is equivalent to a TCDI 50/mL of $10^{7.5}$ on the model used during this experiment. An extract is considered to have antiviral activity when the TCDI 50/mL is lower than the reference value mentioned above. The direct antiviral activity is determined by contact with the virus and the algal extract before reinfection of the cultured cell layer. Cellular protective activity against the virus is established by putting the algal extracts in contact with the splenic cells.

In green algae, *Monostroma nitidum* extract at 10 µg/mL shows a 50/mL TCDI of $10^{5.5}$ (see Table 5.3). It lowers the presence of virus particles by a log 2 compared to infected but untreated control cells. Compared to the other extracts, it showed a higher viricidal activity against HNV. On the other hand, it appears to be less effective against IPNV (TCID 50/mL of 10^7 vs. TCID 50/mL of $10^{7.5}$ for the control). For the other species, viricidal activity is mainly observed for extract concentrations approximately 100 µg/mL. However, extracts from *Ulva pertusa* (green alga) and *Corallina officinalis* (red alga) show efficacy against both viruses.

Algal extract (µg/mL)	HNV viricidal activity (TCID 50/mL)	Viricidal activity VNPI (TCID 50/mL)
Monostroma nitidum (10)	$10^{5.5}$	10^7
Ulva pertusa (100)	$10^{6.3}$	$10^{6.4}$
Sargassum thunbergii (100)	$10^{6.3}$	$10^{6.5}$
Undaria pinnatifida (100)	$10^{6.8}$	$10^{7.1}$
Corallina officinalis (100)	$10^{6.5}$	$10^{6.3}$
Grateloupia filicina (100)	$10^{6.4}$	10^7

Table 5.3. *In vitro viricidal activity of methanolic extracts of green, brown and red algae against infectious hematopoietic necrosis virus (IHNV) and infectious pancreatic necrosis virus (IPNV) (from Kang et al. 2008)*

Independently of the direct viricidal activity, some tested algal extracts proved to be effective in protecting infected cells. This is the case for the red alga *Polysiphonia morrowii* extract which significantly reduces the TCID 50/mL for HNV and IPNV (see Figure 5.12) and thus the level of viral contamination.

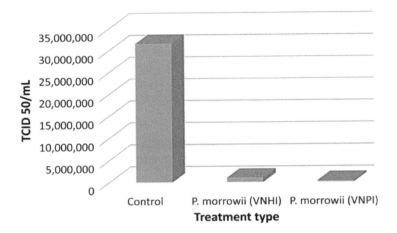

Figure 5.12. *Decrease in viral load due to HNV and IPNV viral loads in splenic cells directly put in contact with an extract of the red alga* Polysiphonia morrowii *(from Kang et al. 2008)*

This observation indicates the presence of compounds in the tested seaweed extracts, reinforcing the cellular protection mechanisms against the studied viruses. These compounds could be sulfated polysaccharides, macromolecules often described for their antiviral properties against human viruses (herpes virus, HIV, Cytomegalovirus) or animal viruses (Kang et al. 2008).

Algal extracts represent a source of active compounds against certain parasites or pathogens targeting plant or animal productions. These compounds have mainly antibacterial or antiviral activities and are of polysaccharide or polyphenolic nature. These properties justify the use of algae as an input in animal feed (see Tables 4.7 and 4.18). In this way, seaweed extracts behave as prebiotics which are useful for the preservation of animal health (see sections 4.2 and 4.4).

Algal extracts also play an important role in plant health. They are involved in the response of plants to abiotic (see sections 3.2–3.4) and biotic stresses. The activity of algal extracts or molecules can be exercised directly on the pathogen (bacteria, virus) or the parasite (nematode). This having been said, the mode of action of algal extracts is mostly indirect. In plants, algal extracts or molecules often act as stimulators of natural defense mechanisms.

In animals, the same molecules, mainly polysaccharides, induce the stimulation of the immune system.

5.2. Induction of plant defense mechanisms

Faced with attacks by parasites or pathogens, plants are likely to develop multiple and varied defense mechanisms. These mechanisms may be biochemical (activation of metabolic pathways or synthesis of bioactive substances) or physical (thickening of the cell wall limiting pathogen penetration).

These defense mechanisms contribute to the notion of "plant immunity", which is well documented in plant pathology.

5.2.1. *The hypersensitivity reaction*

The hypersensitivity reaction is part of the plant immunity process. It is based on the presence of a memory effect that activates a series of defense reactions when the plant is put in contact a second time with a pathogen. The rapidity of the implementation of this defense process is at the origin of the term "hypersensitivity", which qualifies this resistance reaction.

The hypersensitivity reaction is characterized by the appearance of leaf necroses that limit the penetration of the pathogen to the necrotic areas, preventing the pathogen from having a systemic diffusion at the plant level (see Figure 5.13). This defense action is a response that can be activated regardless of the pathogen, bacterium, virus or fungus.

Biochemically, the hypersensitivity reaction results in the activation of enzymes of the phenylpropanoid and phenolamide synthesis pathways (cinnamoyltyramine, coumaroyltyramine, feruloyltyramine) (Fleurence and Negrel 1989). Metabolic activation of these pathways results in the neosynthesis of lignin-like polymers that contribute to cell wall thickening (Fleurence 1988).

It also leads to the activation of numerous glucanase or chitinase-type enzymes that are capable of degrading bacterial or fungal walls. These enzymes activated during the defense mechanism are also known as PR proteins (pathogen-related proteins) (Stintzi et al. 1993). Finally, the hypersensitivity reaction also induces the production of phytoalexins, substances with antimicrobial activities.

Algal extracts and some algal molecules (laminarin, fucoidans) are described as elicitors of plant defense mechanisms and more particularly as inducers of hypersensitivity reactions.

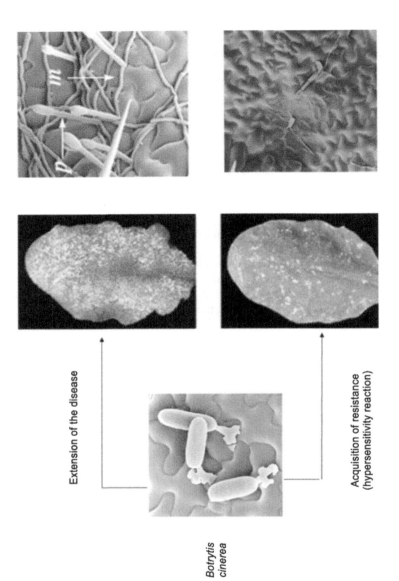

Figure 5.13. *Hypersensitivity reaction resistance mechanism in Arabidopsis thaliana responding to fungal attack by Botrytis cinerea (m: mycelium; p: phialid) (photo credit © C. Veronesi, 2022). For a color version of this figure, see www.iste.co.uk/fleurence/algae.zip*

This is the case for laminarin (see Figure 5.14), a polysaccharide found in the alga *Laminaria digitata*, which activates the hypersensitivity reaction of tobacco against tobacco mosaic virus (TMV). This polysaccharide administered by foliar infiltration regulates the expression of genes involved in the phenylpropanoid pathway, a major metabolic pathway activated during the establishment of plant defense mechanisms (Shukla et al. 2021).

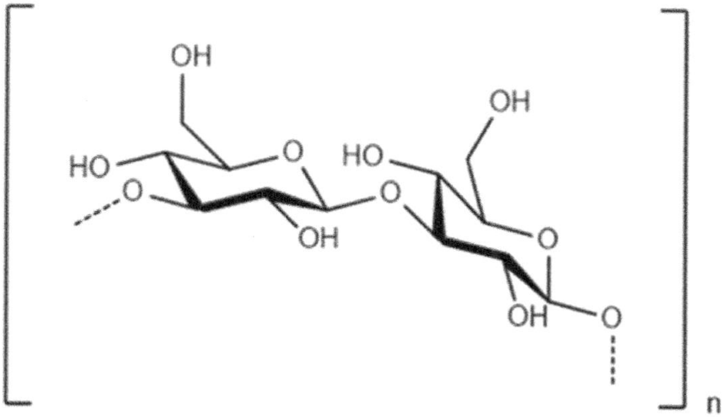

Figure 5.14. *Structure of laminarin*

This molecule is also described as inducing the hypersensitivity reaction of grapevine (*Vitis vinifera*) to the fungus *Plasmopara viticola* (Trouvelot et al. 2008). Its supply strongly limits the progression of the fungus at the leaf level (see Figure 5.15). It also induces the production of H_2O_2 in infected cells and activates the regulation of resistance genes.

The hypersensitivity reaction is a defense reaction based on a memory effect acquired by the plant. It is generally activated during a second contact with the pathogen (acquired resistance). In the absence of a first contact, it can also be induced by the administration of elicitors (induced resistance). These inducing molecules are mainly polysaccharides that mimic the presence of glucans contained in bacterial or fungal walls. The use of these elicitors, most of which are derived from brown algae, is therefore similar to vaccination and appears to be an alternative to the use of pesticides.

This vaccination process, which increases the immune status of the treated plants, however, has a disadvantage. Indeed, under the induction of elicitors, the plants will mobilize part of their metabolism in biochemical defense reactions, and this in the absence of a parasite attack.

In some species, the synthesis of PR proteins can represent up to 10% of the plant protein pool. Under these conditions, preventive treatment with algal elicitors can lead to yield reductions, which explains why this approach is far from being generalized in agriculture.

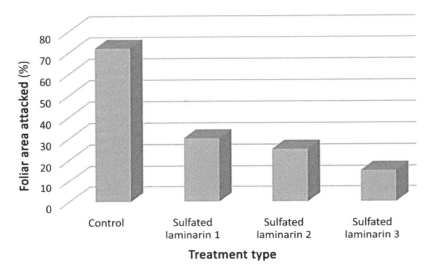

Figure 5.15. Contribution of sulfated laminarin (1: 1.25 mg/mL; 2: 2.5 mg/mL; 3: 3 mg/mL) to the limitation of the foliar progression of the pathogenic fungus Plasmopara viticola *on grapevine (Vitis vinifera) (from Trouvelot et al. 2008)*

5.2.2. *Other mechanisms*

The hypersensitivity reaction is a complex process that is characterized mainly by the confinement of the pathogen (virus, bacteria, fungus) in necrotic areas.

Other defense mechanisms based on the activation of hydrolytic enzymes and the production of bioactive molecules (salicylic acid, H_2O_2, phytoalexins) can also be engaged by the plant in the process of acquired or induced immunity. Inducible mechanisms can be generated, as before, via the administration of algal extracts or molecules (see Table 5.4). These molecules are essentially polysaccharides obtained from brown, red or green algae. Ulvans, sulfated polysaccharides from green algae, of the genus *Ulva,* are among them (see Figure 5.16).

Elicitor	Algal species	Type of administration	Mechanism of action
Alginates	– *Fucus spiralis* – *Bifurcaria bifurcata*	Soaking of the roots	Stimulation of the natural defenses of the roots of date palms
Alginates oligosaccharides	*Macrocystis pyrifera*	Impregnation of cotyledons	Phytoalexin accumulation and phenylalanine lyase (PAL) activation in soybean cotyledons
Laminarins	*Laminaria digitata*	Foliar spraying	Reduction of infection by *Botrytis cinerea*, *Sphaerotheca macularis*, *Mycosphaerella fragariae* in strawberry
Ulvanes	*Ulva fasciata*	Foliar spraying	Induction of resistance in beans to anthracnose caused by *Colletotrichum lindemuthianum*
Ulvanes	*Ulva armoricana*	Infiltration or foliar spray	Activation of plant immunity via regulation of the jasmonic acid synthesis pathway
Oligoulvanes	*Ulva lactuca*	Foliar infiltration	Reduction of the wilting phenomenon caused by *Fusarium oxysporum* infection in tomato
λ-carrageenans	– *Gigartina acicularis* – *Gigartina pistillata*	Foliar infiltration	Induction of resistance in tobacco to *Phytophthora parasitica* via regulation of resistance genes
κ-carrageenans	*Kappaphycus alvarezii*	Foliar spraying	Reduction of anthracnose caused by *Colletotrichum gloeosporioides* in bell pepper by the activation of antioxidant defense mechanisms

Table 5.4. *Examples of algal polysaccharides inducing the resistance mechanisms of some plants of agronomic interest (from Shukla et al. 2021)*

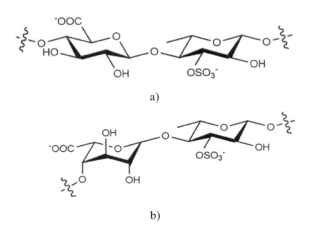

Figure 5.16. Structures of ulvans. a) A ulvanobiouronic acid. b) B ulvanobiouronic acid

The induction of plant resistance to disease has resulted in the development of many commercial products (see Table 5.5). These products are mainly formulated from brown algae and more particularly from *Ascophyllum nodosum*. This species is already frequently used by industrialists for the manufacturing of biostimulant products (see Chapter 3). The base material is composed of aqueous or alkaline extracts.

These products are used to improve plant resistance to parasitic (nematodes, mites), bacterial (*Xanthomonas* sp.) or fungal (*Botrytis* sp., *Alternaria* sp., *Fusarium* sp.) infections. The method of administration is mainly by foliar spraying and more secondarily by root dipping.

Algal extracts possess substances that stimulate the defense mechanisms of plants against most of their pathogens and parasites. In this way, they are suitable to be used in a vaccine approach probably beneficial for plant health.

Although the technical feasibility of such an alternative strategy has been more or less demonstrated, the economic feasibility is still far from certain. Despite this, the use of algal resources for the preservation of plant health remains an interesting societal and economic issue to be evaluated.

Commercial product	Algae	Extract	Types of administration	Plants	Parasites or pathogens	Diseases treated
Algamare	*Ascophyllum nodosum*	Alkaline	Foliar spraying	Japanese plum tree	*Monilinia fructicola* (fungus)	Brown rot
Acadian Seaplants	*Ascophyllum nodosum*	Alkaline	– Foliar spraying – Soaking of the roots	Tomato	– *Alternaria solani* (fungus) – *Xanthomonas campestris* (bacterium)	– Alternaria burns – Bacterial spot disease
Dalgin Active	*Ascophyllum nodosum*	Aqueous	Foliar spraying	– Soft wheat – Durum wheat	*Zymoseptoria tritici* (fungus)	Septoria in wheat
Iodus 40	*Laminaria digitata*	Aqueous	Foliar spraying	Wheat	– *Blumeria graminis* – *Drechslera tritici-repentis* (fungi)	– Powdery mildew – Rhynchosporium
Maxicrop	*Ascophyllum nodosum*	Alkaline	Foliar spraying	Strawberry tree	*Tetranychus urticae* (mite)	Mite infestation
Stimplex	*Ascophyllum nodosum*	Alkaline	Foliar spraying	Cucumber	– *Alternaria cucumerina* – *Botrytis cinerea* – *Fusarium oxysporum* (fungi)	– Alternaria burns – Botrytis blight – Stem rot
Stimplex	*Ascophyllum nodosum*	Alkaline	Foliar spraying	– Tomato – Bell pepper	– *Xanthomonas campestris* (bacterium) – *Alternaria solani* (fungus)	– Bacterial spot disease – Mildew

Table 5.5. *Some examples of algal-based products marketed for their resistance-inducing activities to certain bacterial and fungal diseases and parasitic attacks (from Shukla et al. 2021)*

5.3. Activation of the immune system

5.3.1. *The case of fish raised by aquaculture*

The immune system of fish, like that of mammals, is capable of developing innate and acquired immunity. The innate immunity relies on the action of circulating cells capable of phagocytosing pathogens. Granulocytes, better known as white blood cells, are the main actors of this innate immunity. It also relies on the complement system, which consists of proteins that can form complexes that hydrolyze the membrane of pathogenic cells.

Acquired immunity is an adaptive immunity that is based on a specific response of lymphocyte cells (B lymphocytes, T lymphocytes) to a pathogen. The antibody-producing B lymphocytes are part of the cellular response and the T lymphocytes are part of the humoral response. This immune mechanism is based on a "memory" effect acquired by the organism.

The supply of algae, especially in the form of meal in fish feed, is described to have immunostimulant effects.

This is particularly the case for red sea bream (*Pagrus major*) fed a diet containing 5% *Ulva* meal (Satoh et al. 1987).

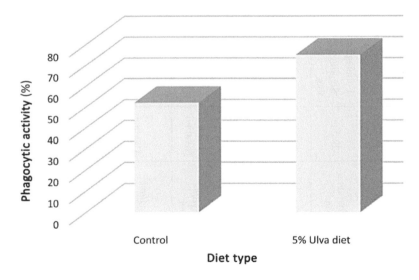

Figure 5.17. *Effect of the addition of Ulva flour on the phagocytic activity of* P. major *granulocytes in contact with the bacterium* Pasteurella piscicida *(from Satoh et al. 1987)*

The addition of *Ulva* flour to the diet of sea bream significantly increases the phagocytic activity of granulocytes (+44%) against *Pasteurella piscicida*, a bacterium that causes septicemia in fish (see Figure 5.17). On the other hand, the number of granulocytes is not impacted by the algae diet. However, this finding is different for lymphocytes whose number increases very strongly following the administration of the algal diet (+13.44%). In fish previously immunized with *P. piscicida* antigen injection, lymphocyte numbers are significantly increased with the intake of the algal diet (see Figure 5.18). This result suggests that the provision of algae in the diet of previously "vaccinated" fish contributes to the enhancement of acquired immunity, thus amplifying the protection of the animals against *P. piscicida*.

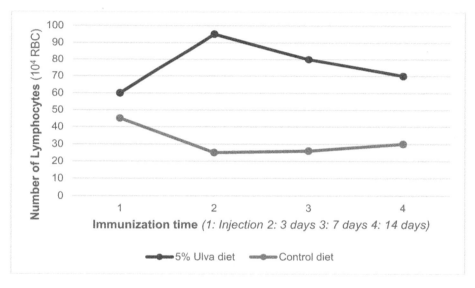

Figure 5.18. *Effect of the Ulva diet on the number of lymphocytes after immunization of* P. major *with* Pasteurella piscicida *antigen (from Satoh et al. 1987). For a color version of this figure, see www.iste.co.uk/fleurence/algae.zip*

Complement activity is estimated by its ability to produce membrane attack complexes and thus to lyse cells. The serum hydrolytic activity of fish is tested in vitro on rabbit erythrocytes. Spontaneous hemolysis of rabbit erythrocytes is higher the stronger the complement activity. Fish fed a diet supplemented with 5% *Ulva* show twice the hemolytic activity of the control group that did not receive the algal feed supplement (see Figure 5.19).

Similarly, the addition of ulva to the diet of fish improves the bactericidal response of fish to various pathogens (*Escherichia coli*, *Pasteurella piscicida*) (Satoh et al. 1987).

The addition of green seaweed in the form of meal at 5% therefore has a positive effect on the immune status of Red Sea bream. This effect concerns both innate immunity (activation of phagocytic activity and the complement system) and acquired immunity (increase in the number of lymphocytes). The integration of algae in the diet of farmed fish to improve their disease resistance capacity is therefore an interesting approach in the case of sea bream, an emblematic species of aquaculture in Southern Europe (Italy, Greece, France, Spain) and Asia (Japan).

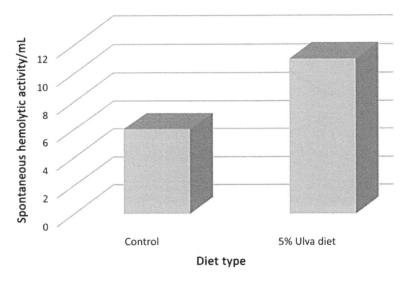

Figure 5.19. *Impact of algal-supplemented diet on the spontaneous hemolytic activity of fish serum (from Satoh et al. 1987)*

Atlantic salmon (*Salmo salar*) is another emblematic species of aquaculture. The incorporation of algal polysaccharides in the feed formulation during the rearing of this species is described as beneficial for the immune status of the animals. The addition of wet feeds incorporating alginates has a positive effect on complement activity (Gabrielsen and Austreng 1998). This results in an increase in spontaneous hemolytic activity and in lysozyme activity, an enzyme involved in the destruction by hydrolysis of bacterial walls.

Stimulation of the innate immune system of salmon is also described with the addition of the red alga *Asparagopsis taxiformis* to the diet of this species (Thépot et al. 2022). The hemolytic activity of fish serum is particularly increased after the application of different diets containing algae or algal extracts incorporated in the diet at a level of about 1%. This activation of the complement pathway is, however, lower than that generated by the supply of lipopolysaccharides obtained from the wall of *Escherichia coli* (see Figure 5.20).

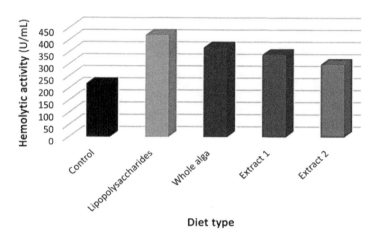

Figure 5.20. *Impact of feeding the alga* Asparagopsis taxiformis *as a whole or as extracts in the salmon diet on the serum hemolytic activity of fish after two weeks of diet (Extract 2 = 2 × Extract 1) (from Thépot et al. 2022)*

The supply of algae also appears to have a positive effect on lysozyme activity, another parameter of innate immunity. An activation of the expression of genes coding for the synthesis of this enzyme is observed after four weeks of treatment (Thépot et al. 2022). In addition, activation of stress response genes, such as HSP70, is also noted after two weeks in fish fed with supplemental whole algae or algal extract (see Figure 5.21).

Algae or algal polysaccharides are often described with regard to their immunostimulatory properties in salmonids. Such properties are also reported for lesser-known species in Europe such as the pearly spot rabbitfish (*Siganus fuscescens*). In this species, the provision of red (*Asparagopsis taxiformis*), green (*Ulva fasciata*) or brown (*Dictyota intermedia*) algae has a positive impact on innate immunity. In particular, this is reflected in a significant increase in serum hemolytic activity in fish supplemented with algae (3% of the diet) (see Figure 5.22).

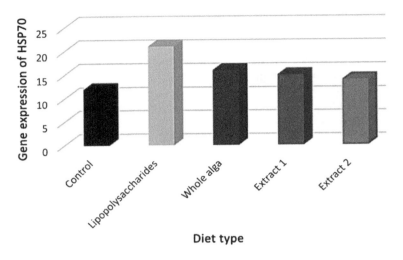

Figure 5.21. *Impact of feeding the alga* Asparagopsis taxiformis *in its whole or extract form in the salmon diet on HSP70 gene expression after two weeks of diet (Extract 2 = 2 × Extract 1) (from Thépot et al. 2022)*

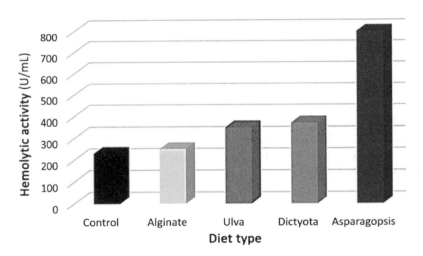

Figure 5.22. *Effect of supplementing green (*U. fasciata*), brown (*D. intermedia*) or red (*A. taxiformis*) algae on the hemolytic activity of rabbitfish (*Siganus fuscescens*) (from Thépot et al. 2021a). For a color version of this figure, see www.iste.co.uk/fleurence/algae.zip*

A meta-analysis of 142 publications shows the impact of algae in the activation of the main mechanisms of innate immunity (hemolytic activity, lysosomal activity, oxidative burst in leukocytes, phagocytic activity) (Thépot et al. 2021b). Three species of aquacultural interest are principally concerned by these publications. These are rainbow trout (*Oncorhynchus mykiss*), Nile tilapia (*Oreochromis niloticus*) and sea bass (*Dicentrarchus labrax*). The algae predominantly tested in these studies belong to the genera *Gracilaria, Hydropuntia* (red algae), *Ulva, Enteromorpha* (green algae) and *Sargassum* (brown algae). Of all the studies concerned, 55 show an activation of lysosomal activity (see Figure 5.23). This increase in activity can be as high as +24.6% when the algae is supplied whole and +36.6% when administered as an extract (Thépot et al. 2021b). The induction of hemolytic activity has been reported in about 30 studies.

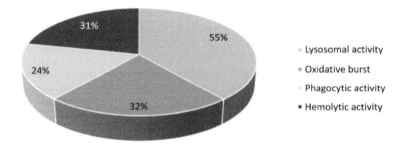

Figure 5.23. *Distribution of innate immunity-related activities by number of studies (142) undergoing meta-analysis (from Thépot et al. 2021b). For a color version of this figure, see www.iste.co.uk/fleurence/algae.zip*

The introduction of marine algae, whether red, green or brown, is described in the scientific literature as having a positive impact on the innate immunity of fish. This induction of the immune response has been shown to be effective against pathogenic bacteria that are particularly dangerous for fish farming. The use of algae is therefore an interesting alternative to the use of antibiotics in aquaculture.

5.3.2. *Other aquaculture animals*

The interest of using algae as an immunostimulant in feed for other aquaculture species has also been studied. These studies are mainly focused on shrimp, the main species of shellfish aquaculture.

The Pacific white shrimp (*Litopenaeus vannamei*) is one of the main species produced by shrimp farming. This species can be affected by the white spot disease caused by WSS (white spot syndrome). The effect of a *Gracilaria verrucosa* (red alga) diet on the immune response of shrimp towards this virus is now well documented (Zahra et al. 2017). The supply of *G. verrucosa* in the shrimp diet induces the innate immune response (hemocyte levels, phagocytic activity, leukocyte oxidative burst, phenoloxidase activity) in infected animals. This response is optimal when the alga is incorporated at a level of 4 g/kg of feed ration. At this concentration, the rate of hemocytes, immune cells with phagocytic activity, is particularly high (see Figure 5.24).

The survival percentage of virus-infected shrimp is also higher in crustaceans fed the diet containing 4 g algae/kg diet. In contrast, a negative effect on shrimp survival was observed when the amount of algae incorporated was increased to 5 g/kg (see Figure 5.25). A dose effect of the incorporation of *G. verrucosa* is thus noted in this kind of experiment.

The inducers of the innate immune response in shrimp are partly known. They are mainly sulfated galactans acting as signal molecules towards hemocytes. Their binding to the membrane of these immune cells is thought to be associated with the activation of shrimp immune response genes (Rudtanatip et al. 2015).

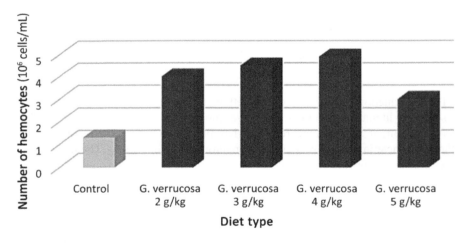

Figure 5.24. *Impact of the dietary intake of* Gracilaria verrucosa *at different concentrations on the hemocyte content in the hemolymph of WSS-infected white shrimp (*Litopenaeus vannamei*) (from Zahra et al. 2017)*

The incorporation of Gracilaria in the diet of shrimps has a zootechnical and health interest. Indeed, these red algae, which can have high protein contents (24% dry weight), contain R-phycoerythrin, a red fluorescent pigment that can color the flesh or the cuticle of crustaceans. That health impact makes these algae a food supplement of choice for shrimp farming.

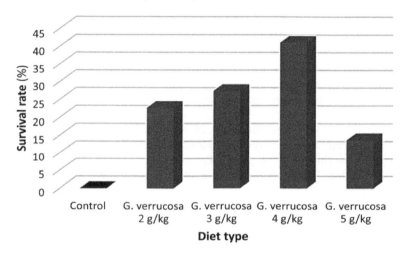

Figure 5.25. *Impact of dietary intake of* Gracilaria verrucosa *at different concentrations on the survival rate of white shrimp infected by WSS (*Litopenaeus vannamei*) (from Zahra et al. 2017)*

5.3.3. The case of terrestrial livestock

Pigs are the most important land animals for production. The world production of pork meat was estimated at 110 million tons in 2020. Numerous studies associate an immunomodulatory activity with algae during pig feeding.

Passive immunity is an acquired immunity transferred naturally from the female to the young via colostrum. This type of immunity has been the subject of numerous studies in pigs. For example, the provision of Laminary extracts during the gestation of sows until the weaning of piglets increases the level of immunoglobulin A (IgA) and G(IgG) in the colostrum by 25% and 19%, respectively (Corino et al. 2019). These extracts contain laminarin and fucoidans, among others. Their administration to the sow (1 g laminarin and 0.8 g fucoidan/day) leads, via lactation, to a significant increase in the IgG level in the serum of the piglet (+10%).

A similar experiment based on the administration of sulfated polysaccharides from the green alga *Ulva armoricana* also shows the impact of these compounds on the passive immunity transferred from sow to piglet. At the end of gestation, sows were fed four different diets, of which one did not contain the algal polysaccharide extract (the control diet). The other three are characterized by polysaccharide administration in increasing doses (2 g/d, 8 g/d, 16 g/d) (Bussy et al. 2019). The presence of antibodies to the *Bordetella bronchiseptica* bacterium, which causes atropic rhinitis in pigs, and their evolution are monitored in sow colostrum and piglet serum. As shown in Figure 5.26, diets with polysaccharide-containing extracts very significantly improve the IgG content of sow colostrum. This increase in IgG content can be observed after 14 days in the serum of piglets suckled by sows fed an algal polysaccharide supplement (see Figure 5.27).

These studies clearly show the effect of the administration of brown or green seaweed extracts enriched in polysaccharides on the passive immunity of pigs. It is interesting to underline that the green alga *Ulva armoricana*, responsible for the green tides on the Breton coasts, represents a resource of natural substances promoting pig immunity. The proliferation of this alga and the ecological disorder caused by it are mainly due to intensive pig farming and the discharge of effluents into the watershed. This double observation leads us to consider a new approach with regard to the valorization of these two fields.

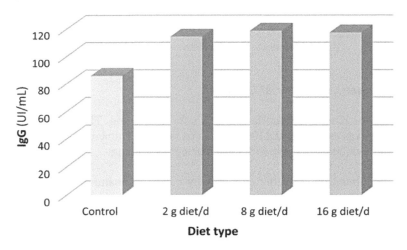

Figure 5.26. *Impact of different diets based on sulfated polysaccharides extracted from the green alga* Ulva armoricana *on the IgG content in sow colostrum (from Bussy et al. 2019)*

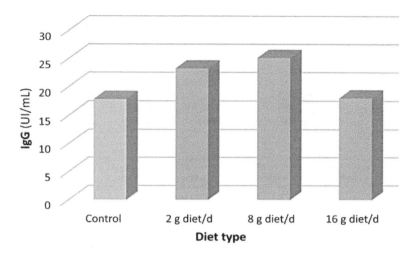

Figure 5.27. *Impact of different diets based on sulfated polysaccharides extracted from the green alga* Ulva armoricana *on the IgG content in piglet serum (from Bussy et al. 2019)*

Passive immunity is linked to lactation and therefore naturally associated with the mammalian class. Apart from pigs, a few studies have investigated the contribution of algal supplementation on this type of immunity in sheep. A study conducted on ewes and their lambs reveals the existence of such a mechanism (Novoa-Garrido et al. 2014). In this study, the provision of a dietary supplement consisting of *Ascophyllum nodosum* (7.2% of the feed ration) shows the presence of an effect on the IgG content in the colostrum after lambing (see Table 5.6). However, no significant effect on IgM content was observed.

Immunoglobulin class	Control diet	Diet supplemented with *Ascophyllum nodosum*
IgG (mg/mL)	75.9	77.6
IgM (mg/mL)	5.7	5.4

Table 5.6. *Effect of supplementing* Ascophyllum nodosum *in the diet of ewes on the content of IgG and IgM in colostrum (from Novoa-Garrido et al. 2014)*

Despite this finding, the algal diet was found to interfere with the transmission of effective passive immunity to *Mycobacterium* sp. Indeed, in the presence of this pathogen, lambs born from ewes fed *A. nodosum* supplementation showed a significantly higher mortality rate (see Figure 5.28) than lambs born from ewes fed another diet, including one with vitamin E supplementation.

168 Algae in Agrobiology

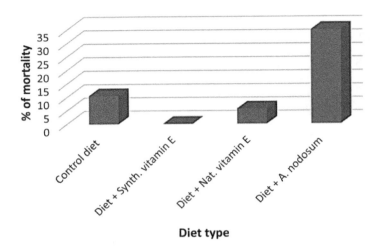

Figure 5.28. *Effect of different diets supplemented with algae, or with vitamin E of synthetic (synth.) or natural (nat.) origin, administered to ewes on the mortality rate of lambs infected with* Mycobacterium *sp. (from Novoa-Garrido et al. 2014)*

This study shows that the transmission of passive immunity remains a complex phenomenon which is very dependent on the animal species and probably on the algal species used, as well as on the form of administration (whole algal meal or extracts enriched in polysaccharides).

In poultry, livestock that are not concerned by this passive immunity, there are a limited number of studies on the impact of algae on the immune status of animals. Among the latter, a recent study describing the contribution of a *Laminaria japonica* extract (3% of the feed ration) on the immune function of broilers should be mentioned (Bai et al. 2019). Animals receiving this dietary supplement had a significantly improved acquired immune response, towards Newcastle disease. This was reflected in an increase in ND virus-specific antibodies and an increase in lymphocyte count (Kulshreshta et al. 2020).

Conclusion

Algae and more particularly macroalgae have always been a resource exploited by humankind to improve the cultivation conditions of plants intended for food. They have supported the development of agriculture, especially in continental coastal regions or islands. Their traditional use was supplanted by the use of chemical fertilizers throughout the 20th century. The reintroduction of algae as a source of organic matter and minerals for the nutrition of cultivated plants appears to be an interesting eco-societal issue for the agricultural and seafood sectors. The biostimulating properties of algae (macro- and microalgae) in terms of plant growth and resistance to biotic and abiotic stresses make it possible to consider an alternative pathway native to the introduction of chemical plant protection products into the soil.

Macro- and microalgae are also used in animal feed, livestock farming being a related activity of agriculture. Macroalgae have traditionally been used for the nutrition of sheep, cattle, pigs and sometimes horses present on the coasts, especially in northern Europe. Their contribution to diet demonstrates the effects on the digestive health of animals. The presence of algal polysaccharides acting as prebiotics plays an important role in the nature of the microbiota of animals. Macroalgae are also interesting food for the breeding of animals produced by aquaculture (mollusks, crustaceans, fish). At a time when many feeds formulated for fish farming include soybeans as the sole source of vegetable protein, it is necessary to question the appropriateness of replacing it with algal protein. This is especially true since soybean cultivation requires a lot of water (900 L of water for 1 kg of soybeans); algae do not require fresh water to grow. Microalgae and cyanobacteria mainly used in shellfish feed are also present in the nutrition of terrestrial livestock (sheep, cattle, rabbits, poultry). However, for these land animals, their use remains marginal, although it is increasing with the development of pet food incorporating spirulina.

Finally, due to their content in bioactive molecules (sulfated polysaccharides, polyphenols), algae are valuable contributors to animal health, limiting the contribution of antibiotic molecules.

Agrobiology, whatever the definition, is an agronomic approach that integrates the development of natural resources while ensuring a sustainable exploitation of the latter and a respect for the environment. The use of algae in agrobiology meets this expectation. Although the approach is old, it was abandoned during the previous century for industrial or economic reasons. The revival of this ancient approach, based on recent research in agrobiology, is another way to promote a sector of plant and animal production closer to our current concerns.

References

Abdel-Twab, S., Saadawy, F., Gomaa, S.A.A., Soliman, A.S., Abbas, M.S. (2020). Effect of foliar application on Arjuna (*Terminalia arjuna*) seedlings under drought and salt stresses. *Plant Archives*, 20, 73–85.

Abubakar, A.R., Ashraf, N., Ashraf, M. (2013). Effect of plant biostimulants on fruit cracking and quality attributes of pomegranate cv. Kadahari kabuli. *Academic Journals*, 44, 2171–2175.

Ali, O., Ramsubhag, A., Jayaraman, J. (2019). Biostimulatory activities of *Ascophyllum nodosum* extract in tomato and sweet pepper crops in tropical environment. *PLOS One*, 14(5), e0216710.

Arieli, A., Sklan, D., Kissil, G. (1993). A note on the nutritive value of *Ulva lactuca* for ruminants. *Animal Production*, 57, 329–331.

Arunkumar, K. and Sivakumar, S.R. (2012). Seasonal influence on bioactivity of seaweeds against plant pathogenic bacteria *Xanthomonas axonopodis* pv. citri (Hasse) Vauterin et al. *African Journal of Microbiology Research*, 6(20), 4324–4331.

Arzel, P. (1994). L'exploitation des algues en Bretagne. *Études rurales*, 133(134), 113–126.

Augris, C. and Berthou, P. (1990). Les gisements de maërl en Bretagne. Report, Ifremer (Centre de Brest), 1–48.

Autret, M. and Madec, A. (2003). Aliment pour poules pondeuses destiné à la production d'œufs comestibles enrichis en iode et en acides gras polyinsaturés et procédés d'alimentation de poules pondeuses correspondantes. European patent application no. 02023843.2

Azaza, M.S., Mensi, F., Ksouri, J., Dhraief, N., Brini, B., Abdelmouleh, A., Kraïem, M. (2008). Growth of Nile tilapia (*Oreochromis niloticus* L.) fed with diet containing graded levels of green algae ulva meal (*Ulva rigida*) reared in geothermal waters of southern Tunisia. *Journal of Applied Ichtyology*, 24, 202–207.

Bai, J., Wang, R., Yan, L., Feng, J. (2019). Co-supplementation of dietary seaweed powder and antibacterial peptides improves broiler growth performance and immune function. *Brazilian Journal of Poultry Science*, 21, 1–10.

Basak, A. (2008). Effect of preharvest treatment with seaweed products, Kelpak and Goëmar BM 86, on fruit quality in apple. *International Journal of Fruit Science*, 8, 1–14.

Battacharyya, D., Babgohari, M.Z., Rathor, P., Prithiviraj, B. (2015). Seaweed extracts as biostimulants in horticulture. *Scienta Horticulturae*, 196, 39–45.

Baweja, P., Kumar, S., Sahoo, D., Levine, I. (2016). Macroalga. Biology of seaweeds. In *Seaweed in Health and Disease Prevention*, Fleurence, J. and Levine, I. (eds). Academic Press, Elsevier, New York.

Béchu, J.Y., Potoky, P., Chasse, C., Le Trividic, D. (1988). Valorisation des algues marines en compostage. In *Valorisation des algues et autres végétaux marins*, Délépine, R., Gaillard, J., Morand, P. (eds). Ifremer-CNRS, Brest.

Becker, E.W. (2013). Microalgae for aquaculture: Nutritional aspect. In *Handbook of Microalgal Culture: Applied Phycology and Biotechnology*, Richmond, A. and Hu, Q. (eds). John Wiley & Sons, Blackwell Publishing, New York.

Belkhodja, M. (1996). Action de la salinité sur les teneurs en proline des organes adultes de trois lignées de fève (*Vicia faba* L.) au cours de leur développement. *Acta Botanica Gallica*, 143, 21–28.

Bensidhoum, L. and Nabti, E. (2021). Role of *Cystoseira mediterranea* extracts (Sauv.) in the alleviation of salt stress adverse effect and enhancement of some *Hordeum vulgare* L. (barley) growth parameters. *SN Applied Sciences*, 3, 116 [Online]. Available at: https://doi.org/10.1007/s42452-020-03992-5.

Blench, B.J.R. (1966). Seaweed and its uses in Jersey agriculture. *The Agricultural History Review*, 14, 122–128.

Blunden, G., Jones, E.M., Passam, H.C. (1978). Effect of post-harvest treatment of fruit and vegetables with cytokinin-active seaweed extracts and kinetin solutions. *Botanica Marina*, XXI, 237–240.

Blunden, G., Jenkins, T., Liu, Y.W. (1996). Enhanced leaf chlorophyll levels in plant treated with seaweed extract. *Journal of Applied Phycology*, 8, 535–543.

Bonomelli, C., Celis, V., Lombardi, G., Martiz, J. (2018). Salt stress effect on avocado (*Persea americana* Mill.) plants with and without seaweed extract (*Ascophyllum nodosum*) application. *Agronomy*, 8, 8050064.

Brundu, G., Monleón, L.V., Vallainc, D., Carboni, S. (2016). Effects of larval diet and metamorphosis cue on survival and growth of sea urchin post-larvae (*Paracentrotus lividus*; Lamarck, 1816). *Aquaculture*, 465, 265–271.

Bussy, F., Salmon, H., Delaval, J., Berri, M., Pi, N.C. (2019). Immunomodulating effect of a seaweed extract from *Ulva armoricana* in pig: Specific IgG and total IgA in colostrum, milk, and blood. *Veterinary and Animal Science*, 7, 100051.

Cabioch, J. (1970). Le maërl des côtes de Bretagne et le problème de sa survie. *Penn ar bed*, 63, 421–429.

Cabioch, J., Floc'h, J.Y., Le Toquin, A., Boudouresque, C.F., Meinesz, A., Verlaque, M. (1992). *Guide des algues des mers d'Europe*. Delachaux et Niestlé, Neuchâtel.

Cardoso, C., Gomes, R., Rato, A., Joaquim, S., Machado, J., Gonçalves, J.F., Paulo Vaz-Pires, P., Magnoni, L., Matias, D., Coelho, I. et al. (2019). Elemental composition and bioaccessibility of farmed oysters (*Crassostrea gigas*) fed different ratios of dietary seaweed and microalgae during broodstock conditioning. *Food Science and Nutrition*. doi:10.1002/fsn3.1044.

Carillo, P., Ciarmiello, L., Woodrow, P., Corrado, G., Chiaiese, P., Rouphael, Y. (2020). Enhancing sustainability by improving plant salt tolerance through macro- and micro-algal biostimulants. *Biology*. doi:103390/biology9090253.

Carmody, N., Goni, O., Langowski, L., O'Connell, S. (2020). Ascophyllum nodosum extract biostimulant processing and its impact on enhancing heat stress tolerance during tomato fruit set. *Frontiers in Plant Science*, 11(807), 1–14.

Cassan, L., Jeannin, I., Lamaze, T., Morot-Gaudry, J.F. (1992). The effect of *Ascophyllum nodosum* extract Goëmar GA 14 on the growth of spinach. *Botanica Marina*, 35, 437–439.

Castilla Gavilan, M. (2018). Diversification de l'activité ostréicole par l'élevage de l'oursin *Paracentrotus lividus*. PhD Thesis, Université de Nantes.

Chanthini, K., Kumar, C.S., Kingsley, S.J. (2012). Antifungal activity of seaweed extracts against phytopathogen *Alternaria solani*. *Journal of Academia and Industrial Research*, 1, 86–90.

Chatzissavvidis, C. and Therios, I. (2014). Role of algae in agriculture. In *Seaweed*, Pomin, V.H (ed.). Nova Science Publishers, Hauppauge.

Chen, J. (2003). Aperçu des méthodes d'aquaculture et de mariculture d'holothuries en Chine. *Bulletin de la CPS*, 18, 18–23.

Corino, C., Modina, S.C., Di Giancamillo, A., Chiappani, S., Rossi, R. (2019). Seaweed in pig nutrition. *Animals*, 9, 1226. doi:103390/ani9121126.

Craigie, J.S. (2011). Seaweed extract stimuli in plant science and agriculture. *Journal of Applied Phycology*, 23, 371–393.

Crouch, I.J. and van Staden, J. (1993). Evidence for the presence of plant growth regulators in commercial seaweed products. *Plant Growth Regulation*, 13, 21–29.

Cruz-Suárez, L.E., León, A., Peña-Rodríguez, A., Rodríguez-Peña, G., Moll, B., Ricque-Marie, D. (2010). Shrimp/Ulva co-culture: A sustainable alternative to diminish the need for artificial feed and improve shrimp quality. *Aquaculture*, 301, 64–68.

Dawes, C. (2016). Macroalgae systematics. In *Seaweed in Health and Disease Prevention*, Fleurence, J. and Levine, I. (eds). Academic Press/Elsevier, New York.

Delaney, A., Frangourdes, K., Li, S.A. (2016). Society and seaweed: Understanding the past and present. In *Seaweed in Health and Disease Prevention*, Fleurence, J. and Levine, I. (eds). Academic Press/Elsevier, New York.

Desouches, M.J. (1972). La récolte du goémon et l'ordonnance de la Marine. *Annales de Bretagne*, 79, 349–371.

Dierick, N., Ovyn, A., De Smet, S. (2009). Effect of feeding intact brown seaweed *Ascophyllum nodosum* on some digestive parameters and on iodine content on edible tissue in pigs. *Journal of Science Food and Agriculture*, 89, 584–594.

Economou, G., Lyra, D., Sotirakoglou, K., Fasseas, K., Taradilis, P. (2007). Stimulating *Orobanche ramosa* seed germination with an *Ascophyllum nodosum* extract. *Phytoparasitica*, 35, 367–375.

El-Deek, A.A. and Brikaa, A.M. (2009). Effect of different levels of seaweed in starter and finisher diets in pellet and mash form on performance and carcass quality of ducks. *International Journal of Poultry Science*, 8, 1014–1021.

El-Iklil, Y., Karrou, M., Mrabet, R., Benichou, M. (2002). Effet du stress salin sur la variation de certains métabolites chez *Lycopersicon esculentum* et *Lycopersicon sheesmanii*. *Canadian Journal of Plant Sciences*, 82, 177–183.

El-Waziry, A., Al-Haidary, A., Okab, A., Samara, E., Abdoun, K. (2015). Effect of strand dietary seaweed (*Ulva lactuca*) supplementation on growth performance of sheep and on *in vitro* gas production kinetics. *Turkish Journal of Veterinary and Animal Sciences*, 39, 81–86.

Evans, F.D. and Critchley, A.T. (2014). Seaweeds for animal production use. *Journal of Applied Phycology*, 26, 891–899.

FAO (2018). La situation mondiale des pêches et de l'aquaculture. Atteindre les objectifs de développement durable. Report, FAO, 2–224.

FAO (2020). La situation mondiale des pêches et de l'aquaculture. La durabilité en action. *Résumé*, 1–27.

Featonby-Smith, B.C. and van Staden, J. (1983). The effect of seaweed concentrate and fertilizer on the growth of *Beta vulgaris*. *Zeitung Pflanzenphysiologie Bd*, 112 S, 155–162.

Férec, C. and Chauvin, T. (1987). L'exploitation des amendements marins dans le golfe normando-breton. *Norois*, 133(135), 229–238.

Fleurence, J. (1988). Purification partielle et étude des propriétés de la tyramine féruloyl transferase extraite des feuilles de *Nicotiana tabacum* variété *Xanthi* n.c inoculées par le virus de la mosaïque du tabac. PhD Thesis, Université de Montpellier II.

Fleurence, J. (2004). Seaweed proteins. In *Technology and Nutrition, Proteins in Food Processing*, Yada, R.Y. (ed.). Woodhead Publishing, Sawston.

Fleurence, J. (2018). *Les algues alimentaires : bilan et perspectives*. Lavoisier, Paris.

Fleurence, J. (2021a). *Microalgae: From Future Food to Cellular Factory*. ISTE Ltd, London, and John Wiley & Sons, New York.

Fleurence, J. (2021b). Perspectives on the use of algae in agriculture and animal production. *Phycology*, 1, 79–82.

Fleurence, J. (2022). Biostimulant potential of seaweed extracts derived from *Laminaria* and *Ascophyllum nodosum*. In *Biostimulants: Exploring Sources and Applications*, Ramawat, N. and Bhardwaj, V. (eds). Springer Nature, Singapore.

Fleurence, J. and Ar Gall, E. (2016). Antiallergic properties. In *Seaweed in Health and Disease Prevention*, Fleurence, J. and Levine, I. (eds). Academic Press, Elsevier, London.

Fleurence, J. and Guéant, J.L. (1999). Algae: A new source of proteins. *Biofutur*, 191, 32–36.

Fleurence, J. and Negrel, J. (1989). Partial purification of tyramine feruloyl transferase from TMV inoculated tobacco leaves. *Phytochemistry*, 28, 733–736.

Fleurence, J., Morançais, M., Dumay, J., Decottignies, P., Turpin, V., Munier, M., Garcia-Bueno, N., Jaouen, P. (2012). What are the prospects for using seaweed in human nutrition and for marine animals raised through aquaculture? *Trends in Food Sciences and Technology*, 27, 57–61.

Frioni, T., Vanderweide, J., Palliotti, A., Tombesi, S., Poni, S., Sabbatini, P. (2021). Foliar vs. soil application of *Ascophyllum nodosum* extracts to improve grapevine water stress tolerance. *Scienta Horticulturae*, 277, 109807.

Gabrielsen, B.O. and Austreng, E. (1998). Growth, product quality and immune status of Atlantic salmon, *Salmo salar* L., fed wet feed with alginate. *Aquaculture Research*, 29, 397–401.

García-Bueno, N., Decottignies, P., Turpin, V., Dumay, J., Paillard, C., Stiger-Pouvreau, V., Kervarec, N., Pouchus, Y.F., Marin-Atucha, A.A., Fleurence, J. (2014). Seasonal antibacterial activity of two red seaweeds. *Palmaria Palmata* and *Grateloupia Turuturu*, on European abalone pathogen *Vibrio harveyi*. *Aquatic Living Resources*, 27, 83–89. doi:10.1051/alr/2014009.

García-Bueno, N., Turpin, V., Cognie, B., Dumay, J., Morançais, M., Amat, M., Pedron, J.M., Marin-Atucha, A.A., Fleurence, J., Decottignies, P. (2016). Can the European abalone *Haliotis tuberculata* survive on an invasive algae? A comparison of the nutritional value of the introduced *Grateloupia turuturu* and the native *Palmaria palmata*, for the commercial European abalone industry. *Journal of Applied Phycology*, 28, 2427–2433. doi:10.1007/s10811-015-0741-z.

García-Vaquero, M. (2019). Seaweed proteins and application in animal feed. In *Novel Proteins for Food. Pharmaceuticals and Agriculture Sources*, Hayes, M. (ed.). Wiley & Sons, New York.

Genovese, G., Faggio, C., Gugliandolo, C., Torre, A., Spanò, A., Morabito, M., Maugeri, T.L. (2012). *In vitro* evaluation of antibacterial activity of *Asparagopsis taxiformis* from the Straits of Messina against pathogens relevant in aquaculture. *Marine Environmental Research*, 73, 1–6.

Gonçalves, B., Morais, M.C., Sequiera, A., Ribeiro, C., Guedes, F., Silva, A.P., Aires, A. (2020). Quality preservation of sweet cherry cv. "staccato" by using glycine-betaine or *Ascophyllum nodosum*. *Food Chemistry*. doi:10.1016/j.food.chem2020.126713.

Goni, O., Quille, P., O'Connell, S. (2018). *Ascophyllum nodosum* extract biostimulants and their role in enhancing tolerance to drought stress in tomato plants. *Plant Physiology and Biochemistry*, 126, 63–73.

Grall, J. and Hall-Spencer, J.M. (2003). Problems facing maerl conservation in Brittany. *Aquatic Conservation Marine and Freshwater Ecosystems*, 13, 55–64.

Guo, Y., Zhao, Z.H., Pan Z.Y., An, L.L., Balasubramanian, B., Liu, W.C. (2020). New insights into the role of dietary marine-derived polysaccharides on productive performance, egg quality, antioxidant capacity, and jejeunal morphology in late-phase laying hens. *Poultry Sciences*, 99, 2100–2107.

Hankins, S.D. and Hockey, H.P. (1990). The effect of a liquid seaweed extract from *Ascophyllum nodosum* (Fucales, Phaeophyta) on the two-spotted red spider mite *Tetranychus urticae*. *Hydrobiologia*, 204, 555–559.

Hansen, H.R., Hector, B.L., Feldmann, J. (2003). A qualitative and quantitative evaluation of the seaweed diet of North Ronaldsay sheep. *Animal Feed Science and Technology*, 105, 21–28.

Haq, T., Khan, F.A., Begum, R., Munshi, A.B. (2011). Bioconversion of drifted seaweed biomass into organic compost collected from the Karachi coast. *Pakistan Journal of Botany*, 43, 3049–3051.

Hegazi, A.M., El-Shraiy, A.M., Ghoname, A.A. (2015). Alleviation of salt stress adverse effect and enhancing phenolic anti-oxidant content of eggplant by seaweed extract. *Gesunde Pflanzen*, 67, 21–31.

Hernandez-Herrera, R.M., Santacruz-Ruvalcaba, F., Ruiz-Lopez, A., Norrie, J., Hernandez-Carmona, G. (2014). Effect of liquid seaweed extracts on growth of tomato seedlings (*Solanum lycopersicum* L.). *Journal of Applied Phycology*, 26, 619–628.

Holden, D. and Ross, R.E. (2012). A commercial extract of the brown seaweed *Ascophyllum nodosum* suppresses avocado thrips and persea mites in field-grown Hass avocados, a pratical field perspective. *Proceedings of the First World Congress on the Use of Biostimulants in Agriculture*, Strasbourg.

Islam, M.M., Ahmed, S.T., Kim, Y.J., Mun, H.S., Yang, C.J. (2014). Effect of sea tangle (*Laminaria japonica*) and charcoal supplementation as alternative to antibiotics on growth performance and meat quality of ducks. *Asian Australasian Journal of Animal Sciences*, 27, 217–224.

Jacobs, R.D., Gordon, M.B.E., Vineyard, K.R., Keowen, M.L., Garza Jr., F., Andrews, F.M. (2020). The effect of a sea-derived calcium supplement on gastric juice pH in the horse. *Journal of Equine Veterinary Science*, 95. doi:10.1016/j.jevs.2020.103265.

Jeannin, I., Lescure, J.C., Morot-Gaudry, J.F. (1991). The effects of aqueous seaweed sprays on the growth of maize. *Botanica Marina*, 34, 469–473.

Jensen, A. (1963). The effect of seaweed carotenoids on egg yolk coloration. *Poultry Sciences*, 42, 912–916.

Jerez-Timaure, N., Sanchez-Hidalgo, M., Pulido, R., Mendoza, J. (2021). Effect of dietary brown seaweed (*Macrocystis pyrifera*) additive on meat quality and nutrient composition of fattening pigs. *Food*, 10, 1720. doi:103390/foods10081720.

Jitesh, M., Wally, O.S.D., Manfield, I., Critchley, A.T., Hiltz, D., Prithiviraj, B. (2012). Analysis of seaweed extract-induced transcriptome leads to identification of negative regulator of salt tolerance in Arabidopsis. *Horticulture Sciences*, 6, 704–709.

Kamel, H.M. (2014). Impact of garlic oil, seaweed extract and imazalil on keeping quality of Valencia orange fruits during cold storage. *Journal of Horticultural Science & Ornamental Plants*, 6, 116–125.

Kamunde, C., Sappal, R., Melegy, T.M. (2019). Brown seaweed (AquaArom) supplementation increases food intake and improves growth, antioxidant status and resistance to temperature stress in Atlantic salmon *Salmo salar*. *PLOS One*. doi:10.1371/journal.pone.0219792.

Kang, S.Y., Kim, S.R., Oh, M.J. (2008). In vitro antiviral activities of Korean marine algae extracts against fish pathogenic infectious hematopoietic necrosis virus and infectious pancreatic necrosis virus. *Food Science and Biotechnology*, 17, 1074–1078.

Kang, S.Y., Kim, M., Shim, C., Bae, S., Jang, S. (2021). Potential of algae-bacteria synergistic effects on vegetable production. *Frontiers in Plant Sciences*, 12, 656662.

Kasim, W., Hamada, E.A.M., Shams, E.D., Eskander, S.K. (2015). Influence of seaweed extracts on the growth, some metabolic activities and yield of wheat grown under drought stress. *International Journal of Agronomy and Agricultural Research*, 7, 173–189.

Kaufman, S., Wolfram, G., Delange, F., Rambeck, W.A. (1998). Iodine supplementation of laying hen feed: A supplementary measure to eliminate iodine deficiency in humans. *Zeitschrift für Emährungwissenschaft*, 37, 289–293.

Khan, W., Hiltz, D., Critchley, A.T., Prithiviraj, B. (2011). Bioassay to detect *Ascophyllum nodosum* extract in *Arabidopsis thaliana*. *Journal of Applied Phycology*, 23, 409–414.

Kofler, L. (1963). *Croissance et développement des plantes*. Gauthier-Villars, France.

Kopta, T., Pavlikova, M., Sekara, A., Pokluda, R., Marsalek, B. (2018). Effect of bacterial-algal biostimulant on the yield and internal quality of Lettuce (*Lactuca sativa* L.) produced for spring and summer crop. *Notulac Botanicae Horti Agrobotanici*, 42, 615–621.

Kulshreshtha, G., Rathgeber, B., Stratton, G., Thomas, N., Evans, F., Critchley, A., Hafting, J., Prithiviraj, B. (2014). Feed supplementation with red seaweeds, *Chondrus crispus* and *Sarcodiotheca gaudichaudii*, affects performance, egg quality, and gut microbiota of layer hens. *Poultry Science*, 93, 2991–3001.

Kulshreshtha, G., Hincke, M.T., Prithiviraj, B., Critchley, A. (2020). A review of the varied uses of macroalgae as dietary supplements in selected poultry with special reference to laying hen and broilers chickens. *Journal of Marine Science and Engineering*, 8, 536. doi:10.3390/jmse8070536.

Kurmaly, K., Jones, D.A., Yule, A.B., East, J. (1989). Comparative analysis of the growth and survival of *Penaeus monodon* (Fabricius) larvae, from protozoea 1 to postlarva 1, on live feeds, artificial diets and on combinations of both. *Aquaculture*, 81, 27–45.

Lami, R. (1941). L'utilisation des végétaux marins des côtes de France. *Revue de botanique appliquée et d'agriculture coloniale*, 243(244), 653–670.

Levigneron, A., Lopez, F., Vansuyt, G., Berthomieu, P., Fourcroy, P., Casse-Delbart, F. (1995). Les plantes face au stress salin. *Cahiers agricultures*, 4, 263–273.

Lombardi, J.V., de Almeida Marques, H.L., Pereira, R.T.L., Barreto, O.J.S., Edison de Paula, E.J. (2006). Cage polyculture of the Pacific white shrimp *Litopenaeus vannamei* and the Philippines seaweed *Kappaphycus alvarezii*. *Aquaculture*, 258, 412–415.

Lopez, C.C., Serio, A., Rossi, C., Mazzarrino, G., Marchetti, S., Castellani, F., Grotta, L., Florentino, F.P., Paparella, A., Martino, G. (2016). Effect of the diet supplementation with *Ascophyllum nodosum* on cow milk composition and microbiota. *Journal of Diary Science*, 99, 6285–6297.

Lucien-Brun, H. (1983). Elevage de juveniles d'ormeaux à fin de repeuplement au Japon. *Publication de l'association pour le développement de l'aquaculture*, 10, 1–54.

MacDonald, J.E., Hacking, J., Weng, Y., Norrie, J. (2012). Root growth of containerized lodgepole pine seedlings in response to *Ascophyllum nodosum* extract application during nursery culture. *Canadian Journal of Plant Sciences*, 92, 1207–1212.

MacKinnon, S.L., Hiltz, D., Ugarte, R., Craft, C.A. (2010). Improved methods of analysis for betaines in *Ascophyllum nodosum* and its commercial seaweed extracts. *Journal of Applied Phycology*, 22, 489–494.

Makkar, H.P.S., Tran, G., Heuzé, V., Giger-Reverdin, S., Lessire, M., Lebas, F., Ankers, P. (2016). Seaweeds for livestock diets: A review. *Animal Feed Science and Technology*, 212, 1–17.

Mannino, G., Campobenedetto, C., Vigliante, I., Contartese, V., Gentile, C., Bertea, C.M. (2020). The application of a plant biostimulant based on seaweed and yeast extract improved tomato fruit development and quality. *Biomolecules*, 10(1662), 1–19.

Michalak, I. and Mahrose, K. (2020). Seaweeds, intact and processed, as a valuable component of poultry feeds. *Journal of Marine Science and Engineering*, 8, 620. doi:10.3390/jmse8080620.

Morais, T., Inacio, A., Coutinho, T., Ministro, M., Cotas, J., Pereira, L., Bahcevandziev, K. (2020). Seaweed potential in animal feed: A review. *Journal of Marine Science and Engineering*, 8, 559. doi:10.3390 /jmse80800559.

Moroney, N.C., O'Grady, M.N., O'Doherty, J.V., Kerry, J.P. (2012). Addition of seaweed (*Laminaria digitata*) extracts containing laminarin and fucoidan to porcine diets: Influence on the quality and shelf-life of fresh pork. *Meat Science*, 92, 423–429.

Muller-Feuga, A. (2013). Microalgae for aquaculture: The current global situation and future trends. In *Handbook of Microalgal Culture: Applied Phycology and Biotechnology*, Richmond, A. and Hu, Q. (eds). John Wiley & Sons, New York.

Mustafa, M.G., Wakamatsu, S., Takeda, T., Umino, T., Nagakawa, H. (1995). Effect of algae as feed additive on growth performance in red sea bream *Pagrus major*. *Fisheries Science*, 61, 25–28.

Norambuena, F., Hermon, K., Skrzypczyk, V., Emery, J.A., Sharon, Y., Beard, A., Turchini, G. (2015). Alage in fish feed: Performances and fatty acid metabolism in juvenile Atlantic salmon. *PLOS One*. doi:10.1371/journal.pone.0124042.

Novoa-Garrido, M., Aanensen, L., Lind, V., Larsen, H.J.S., Jensen, S.K., Govasmark, E., Steinshamn, H. (2014). Immunological effects of feeding macroalgae and various vitamin E supplements in Norwegian white sheep-ewes and their offspring. *Livestock Science*, 167, 126–136.

O'Doherty, J.V., Dillon, S., Figat, S., Callan, J.J., Sweeney, T. (2010). The effects of lactose inclusion and seaweed extract derived from *Laminaria* spp. on performance, digestibility of diet components and microbial populations in newly weaned pigs. *Animal Feed Science and Technology*, 157, 173–180.

Orpin, C.G., Greenwood, Y., Hall, F.J., Paterson, I.W. (1985). The rumen microbiology of seaweed digestion in Orkney sheep. *Journal of Applied Phycology*, 59, 585–596.

Peixoto, M.J., Salas-Leiton Pereira, L.F., Quieroz, A., Magalhaes, F., Pereira, R., Abreu, H., Reis, P.A., Magalhaes Gonçalves, J.F., de Almeida Ozorio, R.O. (2016). Role of dietary seaweed supplementation on growth performance, digestive capacity and immune and stress responsiveness in European sea bass (*Dicentrarchus labrax*). *Aquaculture Reports*, 3, 189–197.

Perez, R. (1997). *Ces algues qui nous entourent. Conception actuelle, rôle dans la biosphère, utilsations, culture*. Ifremer, Plouzané.

Porchas Cornejo, M.A., Cordova, L.M., Magallon Barajas, F., Naranjo Paramo, J., Portillo Clark, G. (1999). Efecto de la macroalga *Caulerpa sertularioides* en el desarollo del camaron *Penaeus californiensis* (Decapoda: Peneidae). *Revista de Biologia Tropical*, 47, 437–442.

Pramanick, B., Brahmachari, K., Ghosh, A., Zodape, S.T. (2014). Effect of seaweed saps on growth and yield improvement of transplanted rice in old alluvial soil of west Bengal. *Bangladesh Journal of Botany*, 43, 53–58.

Quayle, T. (1815). General view of the agriculture of the islands on the coast of Normandy. *Boards of Agriculture*, 148.

Rayirath, P., Benkel, B., Hodges, D.M., Allan-Wojtas, P., MacKinnon, S., Critchley, A.T., Prithiviraj, B. (2009). Lipophilic components of the brown seaweed, *Ascophylum nodosum*, enhance freezing tolerance in *Arabidopsis thaliana*. *Planta*, 233, 135–147.

Rengasamy, K.R., Kulkarni, M.G., Pendota, S.C., van Staden, J. (2016). Enhancing growth, phytochemical constituents and aphid resistance capacity in cabbage with foliar application of eckol – A biologically active phenolic molecule from brown seaweed. *New Biotechnology*, 33, 273–227.

Roque, B.M., Salwen, J.K., Kinley, R., Kebreab, E. (2019). Inclusion of *Asparagopsis armata* in lactating dairy cows diet reduces enteric methane emission by over 50 percent. *Journal of Cleaner Production*, 234, 132–138.

Roque, B.M., Venegas, M., Kinley, R.D., de Nys, R., Duarte, T.L., Yang, X., Kebreab, E. (2021). Red seaweed (*Asparagopsis taxiformis*) supplementation reduces enteric methane by over 80 percent in beef steers. *PLOS One* [Online]. Available at: https://doi.org/10.1371/journal.pone.0247820.

Rossi, R., Vizzarri, F., Chiapparini, S., Ratti., S., Casamassima, D., Palazzo, M., Corino, C. (2020). Effects of dietary levels of brown seaweeds and plant polyphenols on growth and meat quality parameters in growing rabbit. *Meat Science*, 161, 107987.

Rudtanatip, T., Withyachumnarnkul, B., Wongprasert, K. (2015). Sulfated galactans from *Gracilaria fisheri* bind to shrimp haemocyte membrane proteins and stimulate the expression of immune genes. *Fish & Shellfish Immunology*, 47, 231–238.

Ruiz, A.R., Gadicke, P., Andrades, S., Cubillos, R. (2018). Supplementing nursery pigs with seaweed extracts increase final body weight of pigs. *Australian Journal of Veterinary Sciences*, 50, 83–87.

Santaniello, A., Scartazza, A., Gresta, F., Loreti, E., Biasone, A., Di Tommaso, D., Piaggesi, A., Perata, P. (2017). *Ascophyllum nodosum* seaweed extract alleviates drought stress in Arabidopsis by affecting photosynthetic performance and related gene expression. *Frontiers in Plant Science*. doi:103389/fpls.2017.01362.

Satoh, K.I., Nakagawa, H., Kasahara, S. (1987). Effect of the Ulva meal supplementation on disease resistance of Red Sea Bream. *Nippon Suisan Gakkaishi*, 53, 1115–1120.

Sauvageau, C. (ed.) (1920). Utilsation agricole du goémon. In *Utilisation des algues marines*. Librairie Octave Doin, Paris.

Sharma, H.S., Fleming, C., Selby, C., Rao, J.R., Martin, T. (2014). Plant biostimulants: A review on the processing of macroalgae and use of extracts for crop management to reduce abiotic and biotic stresses. *Journal of Applied Phycology*, 26, 465–490.

Shukla, P.S., Borza, T., Critchley, A.T., Hiltz, D., Norrie, J., Prithiviraj, B. (2018). *Ascophyllum nodosum* extract mitigates salinity stress in *Arabidopsis thaliana* by modulating the expression of miRNA involved in stress tolerance and nutrient acquisition. *PLOS One*, 29, 1–25.

Shukla, P.S., Borza, T., Critchley, A.T., Prithiviraj, B. (2021). Seaweed-based compounds and products for sustainable protection against plant pathogens. *Marine Drugs*, 19, 59.

Soppelsa, S., Kelderer, M., Casera, C., Bassi, M., Robatscher, P., Andreotti, C. (2018). Use of biostimulants of organic apple production: Effects on tree growth, yield, and fruit quality at harvest and during storage. *Frontiers in Plant Science*, 9(1342), 1–17.

Stintzi, A., Heitz, T., Prasad, V., Wiedemann-Merdinoglu, S., Kauffmann, S., Geoffroy, P., Fritig, B. (1993). Plant "pathogenesis-related" proteins and their role in defense against pathogens. *Biochimie*, 75, 687–706.

Stirk, W.A. and van Staden, J. (1997). Comparison of cytokinin and auxin-like activity in some commercially used seaweed extracts. *Journal of Applied Phycology*, 8, 503–508.

Strand, A., Herstad, O., Liaan-Jensen, S. (1998). Fucoxanthin metabolites in egg yolks of laying hens. *Comparative Biochemistry and Physiology Part A*, 119, 963–974.

Swarnam, T.P., Velmurugan, A., Lakshmi, N.V., Kavitha, G. (2020). Foliar application of seaweed extract on yield and quality of okra (*Abelmoschus esculentus* L.) grown in a tropical acid soil. *Trends in Biosciences*, 13, 1–6.

Tay, S.A.B., MacLeod, J.K., Palni, L.M.S., Letham, D.S. (1985). Detection of cytokinins in a seaweed extract. *Phytochemistry*, 11, 2611–2614.

Tayyab, U., Novoa-Garrido, M., Roleda, M., Lind, V., Weisbjerg, M.R. (2016). Ruminal and intestinal protein degradability of various species measured *in situ* in dairy cows. *Animal Feed Science and Technology*, 213, 44–54.

Thépot, V., Campbell, A.H., Paul, N.A., Rimmer, M.A. (2021a). Seaweed dietary supplements enhance the innate immune response of the mottled rabbitfish, *Siganus fuscescens*. *Fish & Shellfish Immunology*, 113, 176–184.

Thépot, V., Campbell, A.H., Rimmer, M.A., Paul, N.A. (2021b). Meta-analysis of the use of seaweeds and their extracts as immunostimulants for fish: A systematic review. *Reviews in Aquaculture*, 13(2), 907–933.

Thépot, V., Campbell, A.H., Rimmer, M.A., Jelocnik, M., Johnston, C., Evans, B., Paul, N.A. (2022). Dietary inclusion of the red seaweed *Asparagopsis taxiformis* boosts production, stimulates immune response and modulates gut microbiota in Atlantic salmon, *Salmo salar*. *Aquaculture*, 546, 737286.

Trouvelot, S., Varnier, A.L., Allegre, M., Mercier, L., Baillieul, F., Arnould, C., Daire, X. (2008). A β-1, 3 glucan sulfate induces resistance in grapevine against *Plasmopara viticola* through priming of defense responses, including HR-like cell death. *Molecular Plant-Microbe Interactions*, 21, 232–243.

Vale, B. and Smetana, P. (1965). The effect of seaweed meal on yolk colour. *Journal of the Department of Agriculture, Western Australia*, 6(15).

Valente, L.M.P., Gouveia, A., Rema, P., Matos, J., Gomes, E.F., Pinto, I.S. (2006). Evaluation of three seaweeds *Gracilaria bursa-pastoris*, *Ulva rigida* and *Gracilaria cornea* as dietary ingredients in European sea bass (*Dicentrarchus labrax*) juveniles. *Aquaculture*, 252, 85–91.

Ventura, M.R. and Castanon, J.I.R. (1998). The nutritive value of seaweed (*Ulva lactuca*) for goats. *Small Ruminant Research*, 29, 325–327.

Veronico, P. and Melillo, M.T. (2021). Marine organisms for the sustainable management of plant parasitic nematodes. *Plants*, 10, 369.

Wally, O.S.D., Critchley, A.T., Hiltz, D., Craigie, J.S., Han, X., Zaharia, I., Abrams, S.R., Prithiviraj, B. (2013). Regulation of phytohormone biosynthesis and accumulation in *Arabidopsis* following treatment with commercial extract from the marine macroalga *Ascophyllum nodosum*. *Journal of Plant Growth Regulation*, 32, 324–339.

Wan, A.H.L., Davies, S.J., Soler-Vila, A., Fitzgerald, R., Jonhson, M.P. (2019). Macroalgae as sustainable aquafeed ingredient. *Reviews in Aquaculture*, 11, 458–492.

Williams, A.G., Withers, S., Sutherland, A.D. (2012). The potential of bacteria isolated from ruminal contents of seaweed-eating North Ronaldsay sheep to hydrolyse seaweed components and produce methane by anaerobic digestion *in vitro*. *Microbial Biotechnology*, 6, 45–52.

Williams, T.I., Edgington, S., Owen, A., Gange, A.C. (2021). Evaluating the use of seaweed extracts against root knot nematodes: A meta-analytic approach. *Applied Soil Ecology*, 168, 104170.

Wosnitza, T.M.A. and Barrantes, J.G. (2005). Utilization of seaweed *Ulva* sp. in Paracas bay (Peru): Experimenting with compost. *Journal of Applied Phycology*, 18, 27–31.

Wu, Y., Jenkins, T., Blunden, G., von Mende, N., Hankins, S.D. (1998). Suppression of fecundity of the root-knot nematode, *Meloidogyne javanica*, in monoxenic cultures of *Arabidopsis thaliana* treated with an alkaline extract of *Ascophyllum nodosum*. *Journal of Applied Phycology*, 10, 91–94.

Younis, E.M., Al-Quffail, A.S., Al-Asgah, N.A., Abdel-Warith, A.W.A., Al-Hafedh, Y. (2018). Effects of dietary fish meal replacement by red algae, Gracilaria arcuata, on growth performance and body composition of Nile tilapia *Oreochromis niloticus*. *Saudi Journal of Biological Sciences*, 25, 198–203.

Zafren, S.Y. (1935). Seaweed as fodder. *Trudy Institute Kormov*, 3, 38–86.

Zahra, A., Sukenda, S., Wahjuningrum, D. (2017). Extract of seaweed Gracilaria verrucosa as immunostimulant to controlling white spot disease in Pacific white shrimp *Litopenaeus vannamei*. *Jurnal Akuakultur Indonesia*, 16, 174–183.

Zhang, M., LI, R., Cao, L., Shi, J., Liu, H., Huang, Y. (2014). Algal sludge from Taihu lake can be utilized to create novel PGPR-containing bio-organic fertilizers. *Journal of Environmental Management*, 132, 230–236.

Zmora, O., Grosse, D.J., Zou, N., Samocha, T.M. (2013). Microalga for aquaculture: Practical implications. In *Handbook of Microalgal Culture: Applied Phycology and Biotechnology*, Richmond, A. and Hu, Q. (eds). John Wiley & Sons, New York.

Zodape, S.T., Gupta, A., Bhandari, S.C., Rawat, U.S., Chaudhary, D.R., Eswaran, K., Chikara, J. (2011). Foliar application of seaweed sap as biostimulant for enhancement of yield and quality of tomato (*Lycopersicon esculentum* Mill.). *Journal of Scientific and Industrial Research*, 70, 215–219.

Index

A, B

acid
  abscisic, 32, 43
  aminobutyric acid betaine, 44
  roughanic, 60
Aeromonas hydrophila, 114
Alaria, 74
  esculenta, 8
Algea Valagro, 43
alginate, 27
amendment, 23
Angus cattle, 78
Aphanizomenon, 52
Apostichopus japonicus, 130
AquaArom, 113
Arabidopsis thaliana, 34, 43, 44, 56
Arthrospira platensis, 29, 117
Ascophyllum nodosum, 56
ascorbate peroxydase, 54
Asparagopsis
  armata, 78
  taxiformis, 78
auxins, 23
Azotobacter sp., 30
betaines, 44

biostimulants, 23
Brachionus sp., 116
Brevicoryne brassicae, 140

C, D

canthaxanthin, 103
carcinoculture, 127
carrageenans, 84
Caulerpa sertularioides, 126
Chaetoceros, 118
Chlamydomonas, 53
Chlorella, 53
  vulgaris, 29, 62
Chondria tennusima, 19
Chorda filum, 8
Cocconeis duplex, 121
Corallina officinalis, 137
Crassostrea gigas, 118
Cystoseira, 25
  mediterranea, 50
cytokinins, 23
Cytolan Star, 66
Dasytricha ruminantium, 71
Dicentrachus labrax, 107
Dictyota intermedia, 161

digestive microbiota, 85
ducklings, 99
Dunaliella, 117
   tertiolecta, 124
Durvillaea potatorum, 32

E, F

Ecklonia, 25
   radiata, 92
Entomoneis spp., 111
Eriocheir sinensis, 128
Eucheuma spinosum, 93
Expando, 60
extract
   C129, 59
   PSI-194, 59
fertilization, 11, 23
Fillit, 7
foreshore, 1
fucoidan(s), 27, 84
Fucus serratus, 16

G, H

gelding, 89
glutathione reductase, 54
glycine betaine, 44
Goëmar, 64
   BM86, 63
Gracilaria, 37
   arcuata, 115
   bursa-pastoris, 110
   cornea, 110
Grateloupia turuturu, 124
Grisetang, 8

H

halioculture, 121
Haliotis
   discus hannai, 122
   tuberculata, 124

Himanthalia elongata, 7
Hisikia fusiforme, 97
holothurians, 130
Hordeum vulgare L., 50

I, K

Ictiobus cyprinellus, 114
immunity
   acquired, 158
   innate, 158
Isochrysis affinis galbana, 118
Kappaphycus, 25
   alvarezii, 60
kelp
   April, 7
   cow, 8
   ground, 1
   shoreline, 1
   wrecked, 1
Kelpak, 63, 64
Kutara, 8

L, M

Laminaria
   cloustonii, 7
   hyperborea, 7, 72
   japonica, 99
laminarin, 23, 84
Lithothamnion, 13
Lithothamnium calcareum, 7, 11
Litopenaeus vannamei, 125
lutein, 90
maërl, 7, 11
mantelet, 7
marga, 11
marl, 11
megalop, 128
Meloidogyne
   incognita, 136
   javanica, 138
MYB60, 43

N, O

Nannochloropsis gaditana, 103
Navicula, 121
Nitzschia, 121
OceanFeed Equine, 87
OceanFeed Swine, 80
Oligonychus perseae, 138
Oncorhynchus mykiss, 163
Oreochromis niloticus, 115
Orkney Islands, 69
Oscillospira guilliermondii, 71

P, R

Padina gymnospora, 35
Pagrus major, 158
Palmaria palmata, 8
Paracentrotus lividus, 128
Pasteurella piscicida, 159
Penaeus
 japonicus, 124
 monodon, 124
peroxidase, 54
Persea americana, 45
PGPR (plant growth-promoting rhizobacteria), 27
Phymatolithon calcareum, 89
Pinus contorta, 40
Polysiphonia, 97
Porphyra yezoensis, 104
Portieria hornemannii, 144
Pseudomonas stutzeri, 30
Purina Omolene 100, 89
rambling broomrape, 141
rhizobacterial flora, 67
Rosamin, 111
Rytiphlaea pinastroides, 6

S, T

Sarcodiotheca gaudichaudii, 92
Sargassum
 liebmannii, 35
 sp., 19
Scenedesmus obliquus, 103
Selenomonas ruminantium, 71
semimembranosus muscle, 100
sheep
 in Orkney, 71
 Niamey, 73
Skeletonema costatum, 119
smolts, 113
Solanum
 lycopersicum L., 35, 60
 melongena L., 48
Stoechospermum polypodioides, 138
Streptococcus bovis, 71
Strongylocentrotus droebachiensis, 128
superoxide dismutase, 54
Terminalia arjuna, 52
Tetranychus urticae, 139
Tetraselmis
 chuii, 124
 suecica, 121
thrips, 138
Tilapia aurea, 114
Triticum sativum, 41

U, V

Ulva
 chlathrata, 125
 fasciata, 19
 lactuca, 16
 ohnoi, 111
ulvans, 73, 84
Ulvella lens, 128
Verdemin, 111
Vibrio harveyi, 124
vraic, 3
 growing, 3
 sawn/cut, 3

W, X, Z

wrack, 3
wraec, 3

Xanthomonas axonopodis, 143
zeaxanthin, 90, 103
zoe, 128

Other titles from

in

Agriculture, Food Science and Nutrition

2020

CHERIET Foued, MAUREL Carole, AMADIEU Paul, HANNIN Hervé
Wine Management and Marketing: Opportunities for Companies and Challenges for the Industry

2019

YANNOU-LE BRIS Gwenola, SERHAN Hiam, DUCHAÎNE Sibylle, FERRANDI Jean-Marc, TRYSTRAM Gilles
Ecodesign and Ecoinnovation in the Food Industries

2016

JEANTET Romain, CROGUENNEC Thomas, SCHUCK Pierre, BRULÉ Gérard
Handbook of Food Science and Technology 1: Food Alteration and Food Quality
Handbook of Food Science and Technology 2: Food Process Engineering and Packaging
Handbook of Food Science and Technology 3: Food Biochemistry and Technology

Printed and bound by CPI Group (UK) Ltd, Croydon, CR0 4YY
18/09/2023

08117278-0001